575
RID

Critical acclaim for Mendel's Demon:

'This really is a long-awaited book . . . The book's success is partly because the story is so intrinsically interesting; partly because Ridley knows it inside out; and partly because he tells this story so well . . . read the book yourself; you certainly won't regret it' John Gribbin, *Independent*

'[*Mendel's Demon*] is elegantly written, very informative and free of mathematics . . . Ridley writes like a good lecturer, and he makes the reader feel like a good undergraduate . . . *Mendel's Demon* ends with a very funny riff on the mating habits of angels' *Guardian*

'A brilliant new book, *Mendel's Demon*, by the Oxford evolutionary biologist Mark Ridley (no relation), argues that we are, none the less, about as complicated as life can get . . . he goes on to describe the biology of angels, which are plainly more complex than us and probably therefore have more genes'
Matt Ridley, *Daily Telegraph*

'Ridley's synthesis is a creative act in its own right . . . this succeeds as the first popular account of recent ideas in evolutionary genetics' Sean Nee, *New Scientist*

'*Mendel's Demon* is fiendishly clever, witty and original – an intellectual treat that makes you glad you're not a microbe'
Helena Cronin, author of *The Ant and the Peacock*

'Mark Ridley is one of today's leading evolutionary thinkers. In this book he has identified an important problem and has tackled it with his usual stylish wit, distinctively literate intelligence, and deep knowledge which is both historically grounded and right up to date. He also has a rare knack of being right' Richard Dawkins, author of *The Selfish Gene*

'Mark Ridley is one of Oxford's sharpest thinkers . . . He leads us through the most intriguing reaches of modern biology with literacy and wit. After all, if we are going to disintegrate, we might as well go down smiling' Colin Tudge, *The Times*

'Ridley has much to say that is provoking about how human evolution could yet proceed (if at all) and about how life might have evolved from similar chemical beginnings on other worlds'
New Statesman

Mark Ridley is in the Department of Zoology at the University of Oxford. He was formerly assistant professor of anthropology and biology at Emory University, Atlanta, USA. He has also been a research fellow at St Catharine's College, Cambridge, and at Linacre, Oriel and New Colleges, Oxford. He is the author of several specialist works including *The Problems of Evolution*, *Animal Behaviour* and a highly acclaimed student textbook, *Evolution*, and editor of two anthologies, *Evolution* and *A Darwin Selection*. He is also a frequent contributor to the *New York Times*, the *Sunday Times*, *Nature*, *New Scientist* and the *Times Literary Supplement*, among other publications. He has done research in several areas of evolutionary theory, particularly in relation to animal behaviour.

By Mark Ridley

Mendel's Demon
The Problems of Evolution
Animal Behaviour
Evolution

As editor:
Evolution
A Darwin Selection

Mendel's Demon

Gene Justice and the Complexity of Life

Mark Ridley

A PHOENIX PAPERBACK

First published in Great Britain in 2000
by Weidenfeld & Nicolson
This paperback edition published in 2001
by Phoenix
an imprint of Orion Books Ltd,
Orion House, 5 Upper St Martin's Lane,
London WC2H 9EA

The author and publishers either have sought or are
grateful to the copyright holders for permission
to reproduce the following quotations:
p. 38, six lines from 'Under Which Lyre' from
Collected Poems by W. H. Auden, Faber and Faber, 1994
and Random House, Inc.; p. 260, four lines from
'Damsel in Distress' from *The Complete Lyrics of Ira Gershwin*
courtesy of the Gershwin Estate.

A CIP catalogue record for this book
is available from the British Library.

ISBN 0 75381 410 2

Printed and bound in Great Britain by
The Guernsey Press Co. Ltd, Guernsey, C.I.

Contents

Preface

When the extraterrestrial visitors land on Earth in their space-saucer, they will be excited to see that ours is one of those rare planets where complex life has evolved. They will have found microbes, like our viruses and bacteria, on every life-bearing planet. Naked biological molecules and simple single cells will be so familiar that they will need only some box-ticking on the extraterrestrials' report: 'they are carbon-based; they use DNA; they are fuelled by electromagnetic radiation and chemical resources; they copy themselves.' But only a few planets will have complex life – creatures like us, with our supercomputing brains; or like swans, with their great wings beating out on either side; or like roses, with their symmetric flowers, delicate petals and seductive perfumes. Complex life forms are built from many cells, and each individual grows up from a single cell to an adult that has an organic form and may be capable of intelligent behaviour.

The extraterrestrials will have some experience of complex life, and they will know that the real fun begins when trying to understand complex, rather than simple, life forms. There is no finer cosmic spectacle than the sight of a complex life form struggling to reproduce itself. The first question they will ask is how many copying mistakes Earthly life makes when it reproduces its hereditary molecules. They can then start on the more interesting part of their report. 'The whole lot use DNA, with about fifty catalysts. They make copying mistakes once every thousand million letters or so – which looks about right.' From this, certain consequences obviously follow. 'The big creatures have a thousand million or more letters of DNA code and a hundred or so cell generations in a body. They use sex, of course ...' Other apparent consequences less obviously follow: '... but it's a crazy sort of sex. They pick their partners from only half their kind. The other half are just ruled out ... this is nothing to do with choosing among potential mates, it's some kind of pre-choice inhibition. Anyhow, we'll bring some of the larger ones for the circus. They can go after those music-freaks from the planet near Nemesis, the ones who take in code, one-tenth at a time, from ten different partners, after ten distinct mating dances. The trouble they had keeping the other nine-tenths of their partners' code out is nursery stuff compared with what these Earthlings get up to in egg manufacture.'

*   *   *

Complex life is the masterpiece of nature, and I want to understand it. Life itself poses less of a problem than complex life. The fossil record tells us that life was probably easy to evolve, but complex life was hard to evolve and might not have evolved at all. Simple life appears to have evolved almost instantaneously, as soon as it was possible. But complex life evolved only after a long delay, as if it had been held up by some inherent difficulty. In this book I follow up the idea that the main difficulty was for a complex life form to copy all its genes accurately enough. More DNA is needed to code for a complex life form than for a simple life form. The DNA in a human being is 6600 million letters long and codes for about a hundred thousand genes. In contrast, the DNA of a bacterium is two or three million letters long and codes for two or three thousand genes. Copying mistakes will have become more numerous as the DNA grew longer, for much the same reason as they happen when we are copying written text. A scribe can hope to copy an advertising slogan without making a mistake, and mistakes do not matter much in any case because it is easy to throw the bad copy away and make a new one. But a scribe can hardly hope to copy out the Bible – a job that took a mediaeval scribe about a year and a half – without making some mistakes.

The evolution of complex life required mechanisms to deal with copying mistakes in the DNA. The first mechanisms improved the accuracy of the copying itself. The earliest life forms probably made about one copying mistake in 100 letters, but bacteria had reduced the rate to less than one mistake in 1,000,000,000 letters. This huge improvement is due to the use of DNA for the master-copy – it is an impressively error-proof molecule – and a molecular machinery for proofreading and repairing mistakes. But the possibilities for improving the accuracy of copying seem to have been exhausted by the bacterial stage. The basic DNA copying machinery has remained much the same since then, and we make copying mistakes at a similar rate per letter of DNA as bacteria do. Our total error rate is much higher, however, because we use so many more letters of DNA code. Between bacteria and us the length of the DNA molecule has increased 1000-fold and the DNA has come to be copied 100 times per generation, against the once per generation of a bacterial cell. Our total error rate has gone up 100,000 times, and whereas a bacteria makes a mistake once in every 1000 offspring, we make over 100 mistakes in every offspring. It is something of a paradox how we can persist, while making so many copying mistakes in our DNA. The solution is uncertain, but is probably – sex. Sex can act to concentrate the copying errors in some of a parent's offspring, leaving other offspring relatively error-free. Sexual life forms could evolve to be more complex than clonal life forms.

Sex, however, created a new problem even while it was solving the problem of excess copying mistakes. In a clonal life form, every gene in the parent is put into its offspring: genes are passed on with a probability of 100 per cent. In us, and all complex sexually reproducing life forms, a gene has only a 50 per cent chance of being passed on from a parent to an offspring. The reduction from 100 per cent to 50 per cent in the chance that a gene is passed on was probably a difficult evolutionary step. When each gene had its chance of being passed on cut in half, natural selection would have favoured 'selfish' genes, in one sense of Richard Dawkins' famous expression: genes that could disrupt the system and increase their chances of being passed on to more than 50 per cent. The evolution of complex life was impossible until these selfish genes had been tamed.

The two, related, big themes of this book are both about error. The first kind of error is passive copying error – mutational mistakes in the copying of the DNA. The second kind is active – selfish genes that harm the body by uncooperative and subversive acts. Both kinds of error threaten the existence of complex life. They are also related in their solution. The main solution to the problem of copying error, in complex life, is sexual reproduction. But complex life on Earth uses a particular kind of sex – Mendelian sex: the genes are inherited in the manner first described by Gregor Mendel, in his garden peas at the monastery of St Thomas in Brünn (now Brno, in the Czech Republic) almost 150 years ago. Mendelian inheritance, it turns out, is designed to prevent selfish genes from acts of subversion.

The genes only tolerate, in an evolutionary sense, the reduced chance of being passed on because the lucky genes are picked at random, as if by lot. It is a basic property of Mendelian inheritance that you cannot predict whether or not a particular gene will be passed on. If it could be predicted which genes were to live on in future generations, and which were to die, natural selection could never have brought complex life forms into existence. The genes destined to die would have rebelled and the whole system would have collapsed. God may or may not play dice in the laws of physics and of chemistry. God did not need to play dice in the simple stages of biology, while life reproduced itself clonally. But the evolution of complex life required a mechanism of inheritance with an inherently random component. Somewhere between the bacteria and us – perhaps at about the stage of simple worms – God did have to start to play dice. Life started to use a randomizing system of inheritance, and all subsequent complex life forms have necessarily been built using the randomizing, Mendelian procedure to pass genes from parents to offspring.

Mendelian inheritance controls how genes are inherited in complex life. It combines sex, reproduction, and the probabilistic rather than certain inheritance of genes. Mendel himself was an Augustinian friar, and I like to imagine the chance mechanism as a rather monkish figure – Mendel's demon – who stands over each gene in a parent and decides whether it will be inherited in the next generation, and which other genes it will be passed on with. Mendel was a near-contemporary of the physicist James Clerk Maxwell, after whom the famous (or fairly famous) 'Maxwell's demon' is named. Mendel published his ideas in 1866; Maxwell described his demon five years later. Maxwell's demon is a hypothetical demon. It stands by a hole between two parts of a vessel and, by allowing only the fast-moving molecules through in one direction, can make (without expenditure of work) one part of the vessel hot and the other part cold. Maxwell's demon is an anti-randomizing demon, who opposes the random movement of molecules and produces a more ordered state of the vessel – that is, it comes to have a hot half and a cold half rather than a uniform temperature throughout. Mendel's demon, by contrast, is a more realistic demon. It is a randomizing demon, who creates an ordered state (that is, complex life) by opposing the disruptive force of natural selection.

Complex life depends on Mendel's demon, and complex life

probably did not evolve until the demon was assembled one or two thousand million years ago. Maxwell's demon was a lawless kind of demon, which could violate one of the laws of physics. Mendel's demon is more of a law-enforcing kind of demon. It redirects the laws of biology to a more creative, rather than destructive, direction. There is nothing diabolical about Mendel's demon. People who are familiar with the UNIX computer operating system may see an analogy in the unseen 'daemons' or 'dæmons' that work behind the scenes to perform useful computing tasks. Moral philosophers can think of eudemonism, in which actions are judged by whether they promote happiness. Classicists may think of ευδαιμονια, meaning happiness or good fortune. Pagans may think of the goddess Fortune. Mendel's demon is the executive of gene justice, and we all depend on it for our existence.

The book aims to explain several deep, general features of life. Some of these features, such as sex, are well-known existential puzzles. Others, such as the reproductive cell division called meiosis that determines how genes are passed on, pose puzzles that are just as deep but less well-known. Meiosis is the fateful cell division in which each gene has only a 50 per cent chance of surviving to the end. Meiosis reduces the number of genes by half. So why does it begin not by reducing the number of genes but by doubling them? And why does the demon act only at the first of the two stages during which the gene numbers are reduced? It turns out that the smoke-and-mirrors of meiosis are a trick to baffle the selfish genes. We can therefore understand sex, and the procedures of inheritance, as design features that enabled the evolution of complex life.

It is intellectually satisfying to understand deep, general features of life. It also allows us to predict which features of complex life on Earth will be found in other complex life forms – including future life here, and independently evolved life elsewhere in the Universe. We can predict that all complex life forms will use something like sex and the randomizing procedures of Mendelian inheritance. But other features of Earthly complex life may be more of an accident. Gender is a universal feature of complex life, and intimately related to sex and Mendelian genetics; but its existence may be an accident.

The cells of complex life forms, including ours, contain two independent sets of genes in two different parts of the cell. Mendel's demon keeps one set well behaved, but not the other. The reason why we have these two sets of genes per cell is that complex life on

Earth happened to originate in an accidental merger event, roughly two thousand million years ago. Two cells merged into one, bringing two gene sets into one cell. A cell with two gene sets is likely to suffer the same fate as a merged business that fails to unify its management structures. One of the management teams usually resigns after a business merger, and gender works in much the same way in us. Males eject one of the gene sets from their sperm, but females retain it in their eggs. The long-term consequences have been huge. The sperm have shrivelled in size, as all the potentially troublesome entities have been sucked out; eggs have stayed large, or even grown larger. Sperm, being small, are more abundant than eggs, and this fact underlies the supply-and-demand economics of the mating market. All modern male–female differences (in so far as they are due to evolution) can therefore be traced to an accidental merger event deep in the past. If complex life had not evolved via a merger, it would not have gender. Gender will be the most puzzling feature of complex life on Earth for our extraterrestrial visitors. They will not be able to understand it until they have read our DNA and reconstructed the merger event that is implicit in its codes.

The two big themes of copying error and uncooperative genes make up the core of the book. Chapters 1 and 2 explain the paradox of complex life, the history of complex life on Earth, and the gene number criterion of complexity. Chapters 3 to 5 are about how life has evolved to deal with passive copying error, and Chapters 6 to 8 are about how life has evolved to deal with actively selfish genes. I use the theory of Chapters 3 to 8 to look at human evolution in the present and recent past (Chapter 9) and the possible future evolution of more complex life (Chapter 10). Some of the emerging new reproductive and genetic technologies, such as gene therapy and the preservation of youthful gametes for use later in life, take on a new significance in the grand historical sweep of complex life. They could be the first new error-reduction devices since the evolution of the bacteria 3500 million years ago. Certain kinds of complex life are ruled out at present, because error rates are too high. A reduction in the error rates might allow the evolution of forms with super-human complexity, in intellect or social organization. Alternatively, it might allow the evolution of flexible life forms, that contain the DNA codes for several species but use the code for only one of those species each generation. A flexible life form might contain the code for a fish, a tree, a bird and a human. An individual would, early in life, select which was the best form to grow up as, and switch off the codes for all the other forms. I look at these ideas at the end of the book.

I have added a glossary and end-notes. I have tried to avoid technical terms as much as possible, and defined them when they do appear; but the glossary may help with recurrent technical terms that I have not defined every time I use them. I have also avoided references in the text; they are in the end-notes, together with some provisos and complications that I thought most readers could do without, but some readers might be interested in.

I have been working on the book for an alarmingly long time, and I am grateful to the many people who have helped me, in conversation, in lecture-audience feedback and by e-mail, by telling me about facts, explaining theories, and directing me to references. Alan Grafen and Alex Kondrashov have helped with a particularly large number of my queries. They are high-powered thinkers, and I should say that their help is no kind of imprimatur for what follows. The deleterious mutations are my own. Alan Grafen read two chapters of a much earlier draft, and a residue from them remains in the current Chapter 5. Thanks also to Mark Pagel, who read an early draft of Chapter 3. Big thanks to John Bohannon, Jo Ridley, and my editor at Weidenfeld & Nicolson, Peter Tallack, who read the whole book in its penultimate draft, and helped with comments on style and exposition, as well as the science. I am also grateful to Paul Harvey and Marian Stamp Dawkins, for facilities in the Department of Zoology, Oxford.

Mark Ridley

1
Keeping living things simple

The magnifying Dutchman

Life as we know it (or as most of us know it) is large and complex. Human beings are large and complex, and the familiar kinds of life around us – on the farm, in the garden, around the house – are similar in size and complexity to ourselves. Some familiar life forms may appear to be simple, but even they turn out on investigation to be complex. A butterfly, for example, may look relatively simple, but its internal anatomy is about as complicated as ours. The butterfly's behaviour is simpler than ours, but the way it grows up makes us look positively unimaginative. A butterfly is a meta-morphosed caterpillar, and a caterpillar looks more like a worm than a fluttering, winged imago. A human adult in comparison is little more than an inflated human child. A butterfly has a good claim to be a complex form of life, like all the other life forms we can see around us.

The familiarity of complex life makes it tempting to take the complexity of life for granted. People have been asking why life exists for as long as they have asked questions at all. But it is only recently that people have started to ask why, given that life does exist, it should have evolved to be complex. I shall be looking in this chapter at four reasons why we should not take the complexity of life for granted. Indeed, taken together they almost suggest that it is the complexity of life, rather than the mere existence of life, that is the problem. The first reason is that the familiar creatures around us are a biased sample of all life. Most of the life forms that we share this planet with are tiny, and far simpler than even the simplest creatures we can see with the naked eye. We ultimately owe this discovery – one of the greatest in all human knowledge – to the invention of the microscope.

The light microscope in the late seventeenth century was an even less user-friendly instrument than it is now. It took someone with a unique set of optical and lens-grinding skills to realize its potential: the Dutchman Antoni van Leeuwenhoek (pronounced 'Laywenhook').

The great moment came in 1674, when van Leeuwenhoek visited an inland water, Berkelse Mere, two hours from Delft where he lived. In his words:

I took up a little of it in a glass phial; and examining this water next day, I found floating therein divers earthy particles ... Among these there were, besides, very many little animalcules, whereof some were roundish, while others, a bit bigger, consisted of an oval. On these last I saw two little legs near the head, and two little fins at the hindmost end of the body. Others were somewhat longer than an oval, and these were very slow a-moving, and few in number. These animalcules had divers colours, some being whitish and transparent; others with green and very glittering little scales; others again were green in the middle, and before and behind white; others yet were ashen grey. And the motion of most of these animalcules in the water was so swift, and so various, upwards, downwards, and round about, that 'twas wonderful to see.

The animalcules were probably what naturalists would now recognize as rotifers, ciliates, and *Euglena*. In the next quarter of a century van Leeuwenhoek opened up a new world, in which not only liquid water teemed with little animalcules, but liquid semen with wriggling sperm animalcules, and liquid blood with blood cells. Enlightened observers were delighted. Pope and Swift greeted the new world view in verse, Addison in prose. Certain theological adjustments did, it is true, become necessary; but good religious morals could be drawn. We could be more certain (Addison reasoned) of an ascending hierarchy (angels, gods, and so on) above our sensory capabilities, now that van Leeuwenhoek had revealed the previously unsensed ranks below us. James Thomson was more squeamish. He was grateful to 'the kind art of forming heaven', which had tuned the human sensory system outside the range of these disgusting microbes:

> ... for if the worlds
> In worlds inclosed should on his senses burst,
> From cates ambrosial, and the nectar'd bowl,
> He would abhorrent turn...

You begin to see why W. C. Fields abstained from water.

Louis Pasteur's germ theory multiplied the worlds in worlds enclosed still further in the nineteenth century. The main germs that Pasteur discovered were bacteria. A bacterium is a single cell, and the cell is usually small: much smaller than the little animalcules described by van Leeuwenhoek. Pasteur often had to work indirectly rather than by direct microscopic observation, but his results were still convincing. Only a few eccentrics refused to be (in Bruno Latour's phrase) Pasteurized. Félix-Archimède Pouchet, professor of physiology at Toulouse, explained in 1865 that Pasteur's 'theory of germs is a ridiculous fiction. How do you think that these germs in the air can be numerous enough to develop into all these organic infusions? If that were true, they would be numerous enough to form a thick fog, as dense as iron.' But what would he have made of the next century of microbial research? The microbe hunters who followed Pasteur bagged one bacterium after another. Then came the viruses which are much smaller still. A bacterium is a single cell, and usually so small that you can see it in a light microscope only if you know what you are looking for. But viruses consist of only a few molecules, and are far smaller. They are invisible using a light microscope. They were first suspected in 1883 and definitively isolated in 1935. By that time the electron microscope had been invented (in 1932) and even viruses could be directly observed.

New kinds of microbes continue to be discovered – we shall meet the Archaea, for instance, later in this chapter – but the most striking findings in recent years have been the sheer numbers of microbes, and the astonishing range of environments that they occupy. Viruses and bacteria not only float like an unseen fog in the air, but the soil and seawater are thick with them too. A litre of water, taken from near the sea surface, contains about ten thousand million viral particles and one thousand million bacterial cells. There are immense microbial communities deep in the Earth's crust, beneath the soil on land, and beneath the sediments at the bottom of the oceans. Bacteria lurk deep in oil wells and inhabit solid rock. Even life forms that do not look bacterial contain masses of bacteria. The average human large intestine has a lining of bacteria about 2 centimetres thick, though that is a relatively small amount compared with cows and termites. Our skin is crawling with bacteria. We carry about 1000 to 10,000 bacterial pets per square centimetre of skin, and more like a million per square centimetre in our groins and armpits. A human body is made of about a hundred million million (10^{14}) human cells, together with 10–100 million million (10^{13}–10^{14}) bacterial cells.

3

In a sense, up to half the cells of a human body are bacterial. A group of biologists at the University of Georgia recently estimated just how much of life on Earth is microbial. In terms of numbers of creatures, or numbers of cells, or genetic diversity, the overwhelming majority of life on our planet is single-celled or less, and invisible to us. Bacterial cells are small and their numbers alone might be misleading; but the microbial hegemony is hardly compromised if we use a measure that is independent of the size of the individual creatures. One good measure of the abundance of a life form is the amount of carbon tied up in it in the world today: it turns out that as much Earthly carbon is tied up in bacteria, and related microbes, as in plants. We should imagine, beside the visible biological world of green forests and gigantic trees, of coral reefs and golden prairies of agricultural crops which stretch as far as the eye can see, an invisible microbial world of equal biological mass. Microbiologists have drawn their own conclusions about what God was up to on the seventh day of creation.

Possible worlds

In the year 1600, a philosophical naturalist could hardly have imagined life without complexity. All known life was complex. By 1700 a few simple animalcules had been discovered, but they were still only natural history curiosities. If a philosophical naturalist had asked an existential question about simple and complex life, it would probably have been to question why the simple forms existed at all, rather than to question the large and complex majority. In the next three centuries, our scientific knowledge of simple life forms expanded so much that we now know that life on Earth is dominated by microbes. The large and complex life forms have not been reduced to a curiosity, but they no longer dominate our view of life. Biologists today are tending to ask why complex life exists at all. Why is not all life simple, consisting of single cells, or even molecules without cells?

It is difficult to know which features of life on Earth are inevitable, and which are more accidental. In the evolution of life, as in human history, what happened did happen – and you cannot assess how likely some event was just by looking at it. A question such as whether the history of the twentieth century would have been different without Hitler is easier to ask than to answer, and certainly cannot be answered by observation alone. In fact Hitler lived and did what he did; that is all the evidence we have. Nevertheless, it makes sense that some historical, and evolutionary, events are more inevitable than others. Richard Dawkins has expressed the question in terms of what he calls 'Universal Darwinism'. If life has evolved elsewhere in the Universe, which features of life on Earth will also be found there? In this first chapter I am arguing that it is not at all inevitable that life there would be complex. I should not be surprised to find the planet populated with a similar mass of life as Earth, and with an equally long evolutionary history behind it, but with nothing more complex than single cells. Stephen Jay Gould expressed much the same idea when he asked whether life on Earth would turn out the same if we re-ran the tape of evolution. If evolution happened again, from the origin of life to now, would the planet look much like it is, or different? This is another way of asking how much of the course of evolution was inevitable, and how much accidental. In Gould's terms, if life on Earth re-evolved from its origin until now, I should not be surprised to see the Earth still covered with nothing more complex than bacteria after 4000 million years of evolution.

The meaning of life

How does the non-living matter in the world differ from the living matter? It was historically thought that some special 'vital' force acted to bring matter to life. The idea does survive in Romantic literature – in Mary Shelley's *Frankenstein*, in the iridescent eel-like 'Lamia' of Keats, and the psychedelically bioluminescent sea of Coleridge – but it has been scientifically dead for at least a century. Research moved on to a new kind of definition of life. Biologists in the nineteenth and twentieth centuries sought to define life in terms of attributes that are observable in all living things – attributes such as movement, irritability or metabolism. There are various particular forms of this definition, varying according to the attributes that they identify. The interesting thing about this class of definitions, for our purposes, is that it almost builds complexity into the meaning of life. Attributes such as movement and metabolism are already complicated. However, this whole approach has also now been more or less rejected. One reason is the expanding knowledge of simple life forms, which has made it practically difficult to identify universal attributes of life on Earth. Viruses (or even simpler things called viroids) are the simplest known life forms at present and they have few attributes in common with us. The other reason is more philosophical. None of the attributes, such as movement, are really necessary for something to be alive. The attributes just happen to be present in all the life forms we are aware of. Maybe all life on Earth now, and in the past, has these attributes, but if something evolves in the future, descended from existing life but lacking one of the attributes used to define life, it would still be alive. The definition, not the life form, would be unbiological. We also know about only a minority of all the species that exist, and a species might easily be discovered that lacked one of the attributes used to define life. Again, that would be just too bad for the definition.

If we cannot define life by a vital force, and we cannot define it by universal attributes of known life forms, how can we define it? A better approach is to ignore the attributes of the living creatures we know about and concentrate on the conditions that create life. Life owes its existence to natural selection, and one possible definition of life is anything that can evolve by natural selection. This definition gives us a rather different perspective on the complexity of life, as we can see if we think through the process of natural selection, in order to identify the abstract conditions needed for it to operate.

Natural selection is at work all around us, in all species. When we try to eliminate a disease agent such as the human immunodeficiency virus (HIV) – the agent of acquired immune deficiency syndrome (AIDS) – by means of drugs, we set up a strong selection pressure in favour of viruses that can survive the drug. The viruses evolve drug resistance. They do so because the species contains more than one kind of individual: for example, there are some kinds of HIV that are resistant to drugs, and others that are sensitive. The resistant viruses are more likely to survive and reproduce themselves in the presence of the drug, and their numbers will go up in the next generation. This change in frequency, in which the better adjusted form increases in numbers relative to the other form, is evolution by natural selection.

Natural selection can favour drug resistance in viruses, pesticide resistance in pests, a new camouflage in moths, bigger or smaller beaks in birds, or a change from four-legged to two-legged locomotion in the ancestors of human beings. It can favour a change in any attribute of any living species. However, we need to look behind the individual examples to the abstract workings of the process. Natural selection acts whenever three conditions are met. There has to be variation: that is, more than one kind of individual in the species. The varieties have to differ in how well they survive and reproduce. And the offspring have to resemble their parents: a virus that is drug-resistant has to produce drug-resistant offspring viruses. Any system that satisfies these three conditions will evolve by natural selection. The third requirement – inheritance – is particularly important. If drug-resistant and drug-sensitive viruses were equally likely to produce drug-resistant offspring, evolution would not occur. The drug-resistant viruses might survive better in one generation, but without inheritance this would make no difference to the quantity of drug-resistant viruses in the next generation. Natural selection requires persistent action over the generations, in which one generation builds on the change of the previous generation: and this requires inheritance. If the drug-resistant viruses initially made up 1 per cent of the population, their advantage in the presence of the drug might cause them to increase from 1 per cent to 2 per cent by the time they come to breed. If drug resistance is inherited, 2 per cent would then be the starting point for the next generation, and in the next generation it might increase from 2 per cent to 4 per cent. After a number of generations, the whole species of HIV would have been transformed. But if drug resistance is not inherited, its frequency would never change from the original 1–2 per cent.

In life on Earth, parents resemble their offspring because genes – sequences of DNA (deoxyribonucleic acid) – are physically passed from parent to offspring. (We shall look some more at the workings of genes in Chapter 2.) An offspring virus is not much more than a physical copy of the parental virus's DNA. (HIV actually uses a related genetic molecule, called RNA (ribonucleic acid), rather than DNA. Other viruses use DNA. Viruses are equally able to evolve drug resistance whether they use RNA or DNA.) In a complex life form such as ourselves, the parental DNA is copied and passed on to the offspring via sperm and egg. The inheritance of genes is more complicated in us than in a virus, but inheritance still happens because genes are passed on. The genes are also part of the reason why one individual differs from another within each species. The DNA sequence of a gene may differ slightly between individuals, resulting in differences in the form of these individuals. The genes of some individuals might code for drug resistance, the genes of others for drug sensitivity; which gene becomes more common would depend on whether or not the environment contains drugs.

Natural selection will work on almost anything (such as the DNA or RNA of a virus) that can copy, or replicate, itself. In a population of things that can copy themselves, some versions will probably be able to copy themselves faster than others, and over time those that copy themselves fastest will come to make up most of the population. That is what is happening in a virus that evolves drug resistance: the viral DNA is copied, and DNA that codes for drug resistance is copied more effectively than the DNA that codes for drug sensitivity. Replication alone is not, strictly speaking, enough for natural selection to set to work – the replicating entities also have to vary, and the offspring have to resemble their parents. But a system that can replicate itself will in practice usually satisfy the other conditions. Replication is the main condition for natural selection to operate, and for life to exist.

If life is defined as anything that can evolve by natural selection, what does the definition tell us about what life should look like? It implies simplicity to me. The game is just to copy things, no more. I doubt whether anyone who (unprejudiced by Earthly experience) were to think of inventing a live, replicating system would dream up fantastically complicated beings like us. He or she would invent a simple molecule, capable of copying itself in an environment that contained the simpler constituent chemicals that make up the self-copier molecule. The DNA molecule is of this kind, and a short DNA molecule (or something like one) copying itself over time, perhaps in the sea, is conceptually a much more obvious life form than a fat four-legged animal, with a physiological superstructure of eating and breathing, walking and talking. If the business of life is to copy genes, most of our physiology suggests a loss of focus during evolution; it is not clearly relevant to replication. A hen is only an egg's way of making another egg;* the logic is fine, and (as Jean and Peter Medawar

---

* Generally attributed to Samuel Butler, but Butler implicitly disclaimed its invention, saying 'It has, I believe, been often remarked that a hen ...'.

commented, after quoting the remark in the form 'a chicken is merely the egg's way of making another egg'), 'the only reasonable ground on which one could object to this statement is the pejorative use of "merely", for a chicken is a remarkable and breathtakingly ingenious way of making another egg.' Moreover, even an egg is more egg than DNA, and is an ingenious way of making more DNA. Eggs are huge. We can stay alive eating them. The DNA is a tiny proportion of the egg, the rest of the cumbrous thing having an organization of some complexity, with yolk, albumen, membranes and shell. Once you start to think about how simple something could theoretically be, and be alive, it can seem puzzling that so many of the Earthly life forms consist of fantastically complicated superstructures. Indeed, most of biology consists of studying the superstructures of life, rather than its replicating genetic basis. Open any biology textbook – the college intro- ductory texts now extend to more than a thousand pages – and fewer than ten pages will be about DNA replication; the rest is about bodily machinery that ultimately exists to make that replication possible. It is as if students of literature spent 1 per cent of their time studying the ideas and the literary forms that writers create, and 99 per cent on compositors, typesetters, editors and paper manufacturers. The activities of biologists are never- theless reasonable because of the way life is. Life has evolved complexity way beyond the business of genetic replication. The complicated superstructure interests us not least because when it goes wrong we get ill, but also because it is amazing in itself, as machinery. But why does it exist? Why is Earth not populated by naked, replicating molecules alone?

Test-tube simplicity

People have experimented with simple replicating molecular systems, and the results further dramatize the question. Into a test- tube you pour the raw materials needed for a replicating molecule to copy itself, and you seed the system with replicating molecules of a certain length. Saul Spiegelman and colleagues did some revealing experiments of this kind in the 1960s, at the University of Illinois. They used RNA, not DNA, as the replicating molecule. The initial RNA molecule was a virus called Qβ, which is a chain of RNA. The RNA molecules, like DNA, are chains made up of units called nucleotides. There are four kinds of nucleotide in RNA, symbolized by the letters A, C, G and U. An RNA molecule can be imagined as a sequence of these four letters, something like ... GCGUUA ... (in DNA, another nucleotide called T is used instead of U). Spiegelman's experiments began with a replicating molecule that was slightly over 3000 letters long. It copied itself in the test-tubes, gradually using up the resources, and to keep the system going the experimenters would take a sample of RNA from time to time and transfer it to another test-tube with fresh resources. The experiment is an elemental set-up for evolution by natural selection. Evolution happens because natural selection favours the versions of the RNA molecule that can copy themselves most effectively. The version that is best at copying itself must inevitably increase in frequency over time.

What happened? Did the system evolve more complex molecules – larger replicator molecules that could hog the resources, or suck them up more powerfully, or even eat the smaller replicator molecules? No – what actually happened was that over the generations the replicating molecule shrank. The molecules at the start of the experiment were about 3000 letters long; by the end of the experiment they had shrunk to 550 letters, a sixth of the original length. Natural selection was favouring the molecules that could copy themselves fastest, and the 550-letter molecule was a stripped-down version of the original that could copy itself at maximum speed. The final molecule could copy itself fifteen times faster than the original molecule, six times from the shrinking and another two and a half times because it copied itself faster per unit of length: the letters in the molecule had evolved to match the raw materials better.

The 550-letter molecule may be close to the minimal length for Spiegelman's particular set-up. We cannot be sure, because it was still shrinking when the experiment stopped, but at a much lower rate than earlier on; most of the shrinking had taken place by halfway through the experiment. The future would probably see the molecule either settle down at a length of 550 letters, or shrink a bit more. What did not happen in the experiment, and presumably would not happen no matter how long the experiment were continued, was the evolution of complexity, or anything approaching it. The replicating molecule does not evolve greater length. Rather, it shrinks as it throws off information that has become unnecessary in the cosy conditions of the experimental test-tubes. It does not evolve in the direction of cells, colonies of cells, and monstrosities like you and me. Complexity is not needed and does not happen. Natural selection in Spiegelman's experiment is the survival not of the fattest, but of the thinnest: it is the survival of the simplest, not the most complex. So why did complexity happen in the evolution of real life? What is different about life on Earth from life in the microcosm of Spiegelman's experimental test-tubes?

The calendar of complexity

The dates of the main events in the history of complex life provide a further reason to suspect that the evolution of complexity was improbable. The argument depends on the dates of events that happened in the distant past. Biologists have since the 1980s enjoyed the possession of two independent methods to infer the dates of events in deep time: fossils and molecules. Sometimes the two methods agree; sometimes they do not. The relations between molecular and fossil research – relations that vary from supportive love-in to academic gang-warfare – are one of the great dramas of modern science.

Both methods are easy in principle. To find the date of human origins, for instance, we can look back to the time of the earliest human fossils; humans must have originated before then. Evidence of this kind shows that humans had split from their common ancestor with chimpanzees some time before three and a half million years ago (henceforth abbreviated mya). The other method uses molecular evidence. The DNA molecule seems to change during evolution at a fairly constant rate – a fact (or approximate fact) that is referred to as the 'molecular clock'. The amount of difference between the DNA molecules of two species is therefore proportional to the time back to their common ancestor. Now imagine that we know two species had a common ancestor 10 mya and have accumulated ten differences in a region of their DNA. We can use this as a calibration point to infer the times of the common ancestors of other species pairs, once we have measured the molecular differences for them. If another pair of species have a hundred differences in that DNA region, we can infer they probably split from their common ancestor more like 100 mya. In the case of human origins, the common ancestor of humans and baboons was known to live about 30 mya. Molecular differences were measured between humans and baboons, and between humans and chimpanzees. This allowed a famous inference that humans originated – that is, diverged from the common ancestor of humans and chimpanzees – about 5 mya.

The fossil and molecular methods are both subject to error, sometimes to hefty error. The fossil record is incomplete, particularly for the periods in the distant past, before 550 mya, when many of the big events in the history of live complexity took place (Figure 1). The molecular clock can be an erratic device, and difficult to calibrate; it is also particularly subject to error for dates of more than 550 mya. But although the dates we shall look at are unlikely to be spot on, they are probably reasonable. It is also useful that we have more than one independent method, because our uncertainty is reduced when the two methods agree. Moreover, the main interest of the dates here is for what they tell us about the evolutionary inevitability of complex life, and for this argument the exact dates do not matter. As we shall see, the argument can hold up even if the dates it uses are out by quite a large amount.

We can begin with the origin of life. What is the date of the earliest fossil trace of life? The latest research is for two sites in Greenland; the rocks at both sites contain traces of life. The rocks of one of the sites are over 3700 million years old; those at the other are (more controversially) claimed to be 3800 million years old. The evidence of life consists of chemical traces in the rocks – in particular, of forms of carbon that are probably only produced by living systems. We do not know what life forms produced these carbon traces, though they probably had metabolism of some kind, maybe even photosynthesis. Life at this stage may have been more complex than the very earliest life, which presumably consisted of crudely replicating molecules. It would have taken some time for life with metabolism to evolve from the original molecules, implying a date for the origin of life before 3700 mya. Even a date of 3700 mya is remarkably early. The Earth itself is about 4500 million years old and the planet would have been inhospitable to life for its first few hundred million years; the date is therefore unlikely to be pushed back much more. Life on Earth probably originated approximately 4000 mya. The next interesting

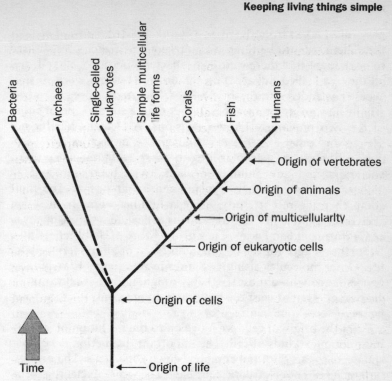

Figure 1. The main events in the evolution of complexity.

date from the fossil record is for the origin of cells. If the life that is fossilized in the 3700-million-year-old Greenland rocks had metabolism, then it was probably already cellular, but the oldest direct evidence of fossil cells comes from Australian rocks called the Apex Chert, and they date back to 3500 mya. There are several finds of fossil cells from Australia at about this stage. Cells are small, and it is difficult to be sure that an apparent fossil cell really is a cell. Any one piece of evidence may prove bogus. However, the total quantity of evidence implies that life existed by this time, and it is generally accepted that cells probably evolved 3500–4000 mya.

What does the molecular clock tell us about the origin of life and of cells? As it turns out, molecular estimates for these early events are either impossible or so uncertain as to be useless. The earliest event anyone has tried to date with the molecular clock is the origin of cellular life. The research makes use of slowly evolving molecules, taken from the most distantly related forms of cellular life. There seem to be three deep divisions of cellular life. It used to be thought that there were two – bacteria on the one hand, and all the rest (called Eukarya) on the other – but the work of Carl Woese has added a third, the Archaea. The Archaea seems to be a wholly new group of microbes, rather than a subdivision of either of the two previously known divisions. Archaeans inhabit hot springs and other extreme environments. A molecular clock estimate of the origin of cellular life can use the difference between a bacterium and any animal or plant, or a bacterium and an archaean, or any animal or plant and an archaean. The most comprehensive recent study, using a number of molecules, has been done by Russell Doolittle and his colleagues at the University of California at San Diego. They give a figure of 'between 3 and 4 billion years ago' for the common ancestor of all modern cellular life. Other molecules also give figures in this range. Clearly these figures are rough – almost laughably rough, with that 1000 million year range – but at least the molecules do not contradict the fossil figure of some time before 3500 mya.

After the origin of cells we move on to the evolution of increasingly complex kinds of cell. The important distinction is between simpler 'prokaryotic' and complex 'eukaryotic' cells. The defining difference is that eukaryotic cells have a separate nucleus, a membrane-bound region inside the cell that contains the DNA. Prokaryotic cells have no nucleus and the DNA lies naked within the cell. The words 'prokaryotic' and 'eukaryotic' are etymologically derived from the Greek *káryon* for a nut, or kernel of a fruit, and by extension a nucleus (*eu*- good, *pro*- before). Eukaryotic cells probably evolved from prokaryotic ancestral cells. This transition was one of the great events – perhaps, as we shall see, *the* great event – in the history of live complexity, because all complex life on Earth is built of eukaryotic cells whereas prokaryotic creatures are all simple. ('Prokaryotes' means the group of all creatures that are built of prokaryotic cells; 'eukaryotes' means the group of all creatures built of eukaryotic cells. No life form contains a mixture of the two cell types.)

Most prokaryotes are microscopic, single-celled creatures. All bacteria are prokaryotic, and for most purposes in this book, 'prokaryote' and 'bacteria' are interchangeable because almost everything that is known about prokaryotes is known from bacteria. It is not really true, however, that all prokaryotes are bacteria: the Archaea are also prokaryotes; but little work has been done on them. We do at least know that they fit in with the general relation between cell type and complexity. All the known archaeans are simple, single-celled creatures, comparable in their complexity to bacteria.

The rest of cellular life is all eukaryotic. The eukaryotes include all kinds of life except the bacteria, Archaea, and non-cellular life such as viruses. Some eukaryotes are single-celled creatures: an amoeba is one eukaryotic cell, as are such ciliates as *Paramecium*, and some kinds of yeast. Practically all multicellular life – and all macroscopic, visible kinds of life – are built from eukaryotic cells. Corals, worms, insects, frogs, birds, us: every cell in every one of these is eukaryotic. Every cell has its DNA in a distinct nucleus (the only exceptions are cells that, like our red blood cells, lack DNA). Seaweeds, moss, conifers and flowering plants are all eukaryotes. These complex kinds of plant and animal life are not only macroscopic and multicellular: they also have individual development. Their bodies contain an internal architecture, with tissues and organs made up of various types of cell: blood cells are different from nerve, bone, muscle or skin cells. The body begins life as a single cell, when the sperm fertilizes the egg, and then develops into an adult with distinct body parts. A few prokaryotes are forms made up of more than one cell, but these are just a mass of similar cells; they do not show development from an egg to a structured adult. A prokaryotic cell is usually just a cell, and stays as the same kind of cell until it divides, buds or dies; a eukaryotic cell may start out as an egg and end up (after some cell divisions) as a brain cell. There are a few exceptional prokaryotes that have more than one type of cell. *Anabaena*, for example, is a common kind of bacterium in ponds and lakes, and each individual *Anabaena* consists of a long chain of cells. Every tenth cell in the chain is a special kind of cell with a distinct biochemical skill: it can 'fix' nitrogen; the nine cells between perform photosynthesis. *Anabaena*, therefore, is both multicellular and has two cell types; but it is still simple compared with a plant or animal, and it is hardly misleading to say that all life forms that have development are eukaryotic. I also think it is hardly misleading to say that all multicellular life is eukaryotic, because the prokaryotic exceptions are so minor. Complex life on Earth is eukaryotic life. What do the fossil and molecular evidence tell us about when this crucial new kind of cell evolved?

The earliest cells to evolve were prokaryotic. The 3500-million-year-old fossils from Australia are probably prokaryotic cells. We have two main reasons to think so. One is that the fossils closely resemble prokaryotes that are alive today. W. J. Schopf has published pictures of the fossils next to similar modern forms, and the similarity is striking even before you allow for 3500 million years of decay. The similarity of the Australian fossils to modern live species illustrates one of the amazing stories of prokaryotic evolution: namely, unevolution – they just seem to stay the same forever. More recent fossil cells are certainly prokaryotes because they are indistinguishable from known prokaryotes that are alive today. Schopf has also published wonderful pictures of some other prokaryotic fossils side by side with modern bacterial species. One species called *Palaeolyngbya*, an almost 1000-million-year-old fossil from Siberia, is pictured beside another species called *Lyngbya*, a modern bacterial citizen of Mexico. I cannot tell them apart. The line from *Palaeolyngbya* to *Lyngbya* has stuck to its simple knitting for 1000 million years, without any need to evolve to become more complex: and the cells that are apparently (if less certainly) fossilized in Australia may have done likewise for the grander period of 3500 million years.

The second piece of evidence is the size of the fossil cells. Prokaryotic cells tend to be smaller: the size range of bacteria is about 0.5–5 micrometres (a micrometre, symbol µm, is one thousandth of a millimetre, and Cyanobacteria, a group of prokaryotes that are common in early fossils, are in the 5–10 µm range. Single-celled eukaryotes range from 5 µm to as much as 2000 µm; most animal and plant cells are in the 10–50 µm range. There is overlap at around 5–10 µm, but sizes are worth knowing about if you are trying to decide whether a fossil cell is prokaryotic or eukaryotic. (Some odd bacterial cells can be much larger. The largest known bacterial cells live in the guts of a species of surgeon fish that swims in the Red Sea: each cell can be 600 µm long.) The 3500-million-year-old Australian fossil cells are about 1.5 µm, well inside the prokaryotic range. Eukaryotic cells evolved some time after that.

The nuclear revolution

The origin of the eukaryotic cell was probably not a single event with a single date, but a series of events spread over hundreds of millions of years. Eukaryotic cells differ from prokaryotic cells in too many respects for it to be likely that they evolved in a single event. I said that prokaryotic cells differ from eukaryotic cells in whether or not the DNA is enclosed in a nucleus, but this is just the formally defining difference from a long list of differences between the two kinds of cell. The insides of a eukaryotic cell are more complex than those of a prokaryotic cell. Eukaryotic cells contain several recognizable substructures called organelles: examples include lysosomes, centrioles and the two structures known as mitochondria (found in almost all eukaryotic cells) and chloroplasts (found in almost all plants). The organelles have vital functions: mitochondria are the organelles of respiration, cellular furnaces where sugar fuels are burned in oxygen to produce the energy currency of the body; chloroplasts are the organelles of photosynthesis, in which energy from sunlight is used to build sugars from water and carbon dioxide. During the origin of the eukaryotic cell, all these organelles had to evolve somehow. Most of them presumably evolved by the standard Darwinian process of modifying a simple precursor until the final structure has been built up. But the mitochondria and chloroplasts almost certainly evolved by a more sudden method. Both these kinds of organelle appear to be descended from free-living bacteria that became incorporated into a larger cell. The cells of modern animals are descended from a merger between two prokaryotic cells, one of which evolved into the mitochondria while the other formed the rest of the cell. The cells of plants have an additional merger in their ancestry, to produce the chloroplasts. Several evolutionary events lie behind the set of organelles in a eukaryotic cell.

The organelles are not the only differences between prokaryotic and eukaryotic cells. Another difference is in their outer covering. Bacteria have a rigid exterior, with a flexible membrane inside. The rigid wall was lost when eukaryotes evolved, and an internal set of props, called the cytoskeleton, evolved to support the cell instead. A eukaryotic cell is surrounded by a permeable lipid, more like a soap bubble than a rigid exterior wall. The eukaryotic cell membrane made it possible for the cell to engulf relatively large prey and digest them internally. Bacteria do not have internal digestion: they nourish themselves by absorbing nutrients from their surroundings; they either secrete enzymes onto large food items and absorb the digest, or absorb small nutrient molecules directly. The evolution of internal digestion, associated with the flexible outer surface membrane of eukaryotes, would have opened up new ways for the cell to make a living, and permitted the evolution of larger cell size. (Some eukaryotes have subsequently re-evolved rigid exteriors, but that is another matter.)

Eukaryotic cells are really into membranes. They have a maze of internal membranes, connecting the outer membrane with the insides of the cell and with the membrane that surrounds the nucleus. The internal membranes act as canals inside the cell, and also keep the molecular assembly lines in order. Molecular messages are sent to and from the DNA in the nucleus via the membrane system. The gates in the nuclear membrane are fantastically complicated structures called 'nuclear pores'. The nuclear pores are the most complex structures inside a cell, consisting of hundreds of molecules. They are presumably a bureaucrat's dream of a customs and immigration department, with each molecule that goes in or out of the nucleus being checked, and probed in its intimate orifices, and pushed about. Bacterial cells use more open government. The DNA lies in the cell and any molecule can reach it by diffusion. All these membranes required further events in the evolution of the eukaryotic cell.

Eukaryotic cells have a distinct method of cellular reproduction. The genes and other cellular components first double up inside the cell. A special machinery of cables forms inside the cell, and they mechanically pull the two sets of genes into the two opposite halves of the parent cell. A membrane then forms between the two halves and division is complete. Such is the normal process of cell division, called mitosis, for instance in a growing animal or plant. During sexual reproduction, eukaryotes use a related process of cell division called meiosis. Meiosis is the mechanism behind the kind of genetic inheritance – the kind first discovered by Gregor Mendel in garden peas – that is only found in eukaryotes. It is the biological mechanism of what I shall call Mendel's demon. No prokaryotes are known to use anything like mitosis or meiosis: they do not spin any machinery of gene-pulling cables. Dividing prokaryotic cells just split, or bud. The machinery of cell division will have required yet further stages in the origin of the eukaryotic cell.

The features of the eukaryotic cell, or some of those features, are the keys to the evolution of complex life on Earth. Ultimately we shall want to know the timing of each stage in the transition from prokaryotic to eukaryotic cells. One of the most important dates would be for the defining stage in the whole transition: the origin of the nucleus. But neither the fossil nor the molecular evidence has been able to tease apart the stages in the evolution of the eukaryotic cell. What we can do is use the molecular clock to date the beginning and the end of the whole process. Doolittle's team gave a figure of 2200 million years for the split between bacteria and eukaryotes; this is the early date, likely to be just before the whole evolution of the eukaryotic cell was begun. They gave a figure of 1500 million years for the common ancestor of the main modern eukaryotic groups; this is the late date, likely to be after the whole process was complete. Both figures are uncertain. In particular, the mergers that produced the mitochondria and chloroplasts can be a nuisance. Mergers have consequences not unlike attaching a second set of hands, pointing in an arbitrary direction, to a real clock face; it becomes difficult to tell the time. The 1500–2200 million-year figures are meant to be corrected for the merger confusion, but the correction may be imperfect.

And what about the fossils? Early fossil eukaryotic cells have been known for some time from the 1600–1000 mya period: but then in 1992 some fossil algal cells (algae are eukaryotes) turned up on the side of a mine pit in Michigan and the eukaryotic fossil record was pushed dramatically back, by 500 million years. It is not certain that the 2100-million-year-old Michigan fossils were eukaryotes. The fossils do not preserve the outlines of individual cells, and we cannot measure cell size for sure. But the fossils look identical to later corkscrew-shaped algae that are known to be single cells. If the Michigan fossils really are identical to the later forms, then we can infer that the fossils are of single cells and their sizes are huge. They are 1 cm (10,000 μm) or so – well into the eukaryote zone. In all, we can be reasonably confident that eukaryotic cells had evolved before 1600 mya, and probably before 2100 mya. An approximate figure of 2000 million years often used to be quoted from the fossil evidence, but that was before the 1992 publication about the Michigan corkscrews. We should now shift the approximate figure back a few hundred million years. The real date may be much earlier. In 1999, an Australian group argued for a date of more like 2700 mya. Their evidence consisted of chemical traces in the rocks – traces of certain fats that may be produced only by eukaryotes. It is difficult to be sure how good a signature these fats provide of eukaryotic, as opposed to pro-karyotic, life – or of what stage of eukaryotic life. It would be premature to build any arguments on the evidence; but it is worth keeping in mind.

How well do the fossil and molecular evidence agree? We do not have a fossil date for the beginning of the evolution of the eukaryotic cell. The Michigan corkscrews were fully evolved euka-ryotes (provided the interpretation of them as algae is correct), and imply that the eukaryotic cell had completely evolved by 2100 mya. That is older than the molecular date. The molecular clock suggests that the process started more like 2200 mya and did not end until some time before 1500 mya. However, both kinds of evidence are uncertain and both point to the same approximate figure, of about 2000 mya. We also have a third kind of evidence, pointing to the same approximate date. The atmospheric con-centration of oxygen, which had been negligibly low since the origin of the planet 4500 mya, suddenly shot up in a spurt about 2000 mya. The connexion between the eukaryotes and the oxygen spurt is uncertain but is unlikely to be a coincidence. The increase may have followed the merger of chloroplasts into eukaryotic cells, which improved the power of photosynthesis. Mitochondria, in turn, probably evolved after the oxygen increase, because they are efficient only when they can burn fuel in oxygen. We therefore have three independent pointers, uncertain individually but fairly convincing in combination, and they all point to a date near 2000 mya for the origin of the eukaryotic cell.

So much for the eukaryotic cell. Our next step is the origin of multicellular life, or multicellular eukaryotic life to be exact. Fungi, plants and animals are three of the main groups of modern many-celled life. Fossils of multicellular life appear in serious amounts in the 'Cambrian explosion', about 540 mya; but there is a site in China that dates from about 580 mya and contains both multi-cellular animals and multicellular plants. (As I write, the German palaeontologist Dolf Seilacher and his colleagues have a controversial claim of 1000-million-year-old fossil traces of worms; but it is not yet reliable.) From fossil evidence, it is safe to conclude that multicellular eukaryotes, including animals, originated some time before 580 mya.

The molecular clock has also been used to date the origin of animals. Several studies agree at about 1000–1200 mya; the Doolittle team I mentioned before finds a more recent figure, 850 mya. A 1996 paper by Gregory Wray and colleagues is one of the most thorough pieces of work supporting a 1200 mya figure: but their date was actually for an event after the origin of the animal branch. The initial origin of animals would have been earlier still. I do not think these molecular dates badly contradict the fossil dates. The fossil forms living 550–600 mya show so much variety that they may well have been evolving apart for a few hundred million years. By 550 mya, we have fossil sponges, sea anemones, worms, arthropods – and they would not have evolved geologically overnight. However, many people think the molecular and fossil dates are in conflict, and some of them distrust the molecular 1000–1200 mya date (or even 850 mya) and think that animals originated nearer the time of the first fossil animals, about 600 mya. That means we have two views about the origin of multicellular life as a whole and of animals in particular. The eukaryotic cell, we saw, originated over 2000 mya. If animals had evolved by 1200 mya, multicellular life must have originated between 2000 and 1200 mya – perhaps about 1500 mya. If animals did not evolve until more like 600 mya then multicellular life originated between 2000 and 600 mya – perhaps more recently than 1500 mya and in the 1500–1000 mya range.

Such are the main dates in the history of live complexity. They give an idea of when the events happened, rather than being exact figures. Perhaps the main thing to stress is that the knowledge is in the grey zone, where the evidence is so poor that we are on the edge of ignorance. Maybe we are peering into total darkness. We need more fossils and better molecular clocks. But the dates we have are not worthless, and I have summarized them in Table 1. I have added two more figures, for the origin of vertebrates and of humans, that will be useful later. (Vertebrates are the group that consists of fish, amphibians, reptiles, birds and mammals; they have backbones and brains that are related to ours.)

Table 1. Time of main events in the increasing complexity of life, estimated from fossil and molecular evidence.

Event	Time (millions of years ago)	
	Fossil estimate	Molecular estimate
Origin of:		
life	3700+	
cells	about 3500	3000–4000
eukaryotic cells	2200–1600	2000
multicellular life/development	[1000–1500]	
animals	600+	1000–1200
vertebrates	500	590–700
humans	5	5

The Rockefeller and the shoeshine boy

The dates in Table 1 arguably tell us something about the probability of the evolutionary events, including the probability of complex life. On a human timescale we can infer how probable events are from how often they happen, or from how long it is since one last happened. The insurance industry uses probabilities in this sense. Motor accidents are highly probable; they happen every day, and the most recent one probably happened round about when you were reading the previous word to this. Natural disasters that do more than US $10 billion of damage are less probable; they are rare but not uncommon, happening every few years, maybe twice a decade – the Kobe earthquake was the most recent as I write. Natural disasters are less probable than motor accidents. My understanding of probability here is that it refers to the number of causal preconditions needed and how likely they are to be together in the right place at the right time, given the way the world is. If one kind of event (such as motor accidents) requires three causal preconditions – perhaps (a) one active motorist, (b) another active motorist or a tree, (c) carelessness – and all three are abundant in the world and frequently found together, that event is probable. If another event requires 300 causal pre-conditions and some of these factors are rare in the world, or mutually incompatible most of the time, then the event is improbable. An event may be improbable because it requires many independent preconditions, or a few that are rare or rarely found together, or some compound of the quantity, frequency and coincidence of its preconditions.

A similar kind of reasoning can be used with evolutionary events. Life seems to have originated almost as soon as it could, implying that the origin of life is a very probable event given the way the world is – or the way it *was*, 4000 mya. This is an important respect in which our modern view of evolution differs from the 1950s and 1960s. Biologists then liked to dramatize the improbability of the origin of life (and its difficulty as a research problem) by arguing that the origin of life was an evolutionary event of almost unrivalled difficulty. The origin of life, they argued, was so improbable that it may have been unique, requiring such a fluke, or conditions so peculiar, that it could never be repeated in the laboratory. Jacques Monod's famous book *Chance and Necessity* includes this argument. The argument is interestingly falsifiable, as Karl Popper remarked; it would be refuted if someone managed to resynthesize a living system. But the dates of the earliest fossil evidence of life have been staked further and further back since Monod's book, and they are now near the earliest possible limit. The old view has lost its appeal. Francis Crick, in his book *Life Itself*, did find a way to stick with the old view (he was betraying his years) in the face of the fossil evidence: he argued that life was introduced to Earth from space. That is another, reasonable interpretation of the evidence, but the inference that most people make from the dates is that the origin of life is easy. We are left only with the puzzle of why laboratory research has failed to recreate it. I suspect it will prove to be one of those problems that continues to look difficult all the time until someone sees how to do it, when it will immediately look easy. Before 1950, it appeared to be difficult to create the biochemical building blocks of life, such as amino acids, sugars and nucleotides, from elementary chemicals such as water, nitrogen and carbon dioxide. But then Stanley Miller found the experimental conditions that were needed, and the whole problem was relegated to a schoolroom exercise. Who would be surprised if the laboratory synthesis of a replicating molecular system underwent a similar fate in the next ten years.

The dates imply that the origin of life is easy. What about the origin of complex life? Here they hint at a different answer. Complex life, in the form of, say, multicellular creatures, did not evolve until 1000–1500 mya, and fish (the first vertebrates) did not evolve until 500 mya. Life existed exclusively in the relative simplicity of single cells, reproducing themselves down the generations, for 2000 million years, a longer period of Earth's history than the subsequent period with complex multicellular life. This implies there is something not so probable about the evolution of complexity. The puzzle is not life, it is complex life.

The inference that life is probable, and complex life improbable, is reasonably good. The dates so overwhelmingly point to it that we end up with the same conclusion no matter how we conjure with the evidence. The Earth is 4500 million years old, and life had probably originated by 4000 mya. Only 10 per cent or so of the history (up to now) of the planet was prebiological, and even that 10 per cent was probably a phase when life on Earth was impossible. True, the date of the origin of life is uncertain, but even if the 3700-million-year-old evidence from Greenland is exploded, a figure of well over 3500 million years would still be well supported: the prebiological phase cannot be stretched beyond 20 per cent of the planet's history. Now turn to complex life. Multicellular life is unlikely to have existed much before 1500 mya, nor animals much before 1200 mya. Indeed the main controversy about these numbers is whether animals may not have proliferated until much later, maybe 600 mya. Complex life is but a terminal curiosity in the great sweep of Earthly history, being tacked on to the last 15–20 per cent of Earth's 4500-million-year existence. The 4500 million years are made up of an initial 10 per cent phase before life, then a whopping 70–75 per cent phase of simple life (made up of less than a cell, or one cell, or maybe at most a few cells), and finally a 15–20 per cent phase when complex life evolves. Complex life looks evolutionarily difficult.

I have started talking about 'easy' and 'difficult' evolutionary events. These words are really no more than metaphors for 'probable' and 'improbable' on the same concept of probability as before (where the probability of an event depends on the number, frequency and interaction of its causal preconditions). I emphatically do not mean that life was trying to evolve complexity, and having difficulties despite heroic efforts. The dates indicate that the factors needed for the origin of life were abundantly present, but that the evolution of complex life may have required some awkward coincidences or rare causal factors. The causal factors themselves are mindless, and operate under influences unrelated to whether or not they will lead to complex life in the future. It just so happens that if they do coincide, then Earth becomes populated with complicated living creatures.

Can we use the dates to identify what the improbable step was (or steps were) that held up the evolution of complex life? The uncertainties in the dates are particularly frustrating here, but the dates in Table 1 make it tempting to guess that the difficult, and crucial, step was either the origin of the eukaryotic cell or the origin of multicellular life. Bacteria – that is, prokaryotic cells – evolved soon after the origin of life: the evolution of cellular life, therefore, may be an easy step. After that, the range of dates allows two extreme lines of reasoning. We could accept an early date for the origin of single-celled eukaryotes, at more than 2700 mya, and a late date for the origin of many-celled creatures, at 600 mya. Then the history of life sees a rapid origin of cells, a relatively rapid origin of eukaryotic cells, followed by a long delay, of about half the history of life, before multicellular life evolves. The difficult step would look like the evolution of multicellularity.

Alternatively, we could accept a later figure for the evolution of a complete eukaryotic cell, at 2000 mya, and an early evolution of multicellularity, by 1500 mya. Then, after the origin of cells, about half the rest of the history of life may have passed before eukaryotic cells evolved and multicellular life evolved relatively quickly after that. The difficult step would look like the evolution of the eukaryotic cell. We could also conjure with the dates in a less extreme manner, and argue that the history of life contains two difficult steps – from prokaryotic cell to eukaryotic cell, and from single-celled to many-celled eukaryotic life.

The inference, or guess, about which step was improbable within the history of complex life is flimsy compared with the inference about the relative evolutionary probabilities of life and of complex life. The dates are less reliable. The history of the planet has been dominated by single-celled life, however you read the evidence. But the dates for the origin of eukaryotes and of multicellular life are highly uncertain, and the evolution of the eukaryotic cell could itself have been a long process, lasting hundreds of millions of years. We can draw a relatively confident conclusion about the improbability of complex life, but only a weak conclusion about what it was that made complex life improbable.

The arguments we have been looking at are 'reverse probability' arguments, in which we infer the chance that something will happen from the time it takes. The argument is stuffed with uncertainties: the dates themselves are probably inaccurate, and the times are for unique events, rather than frequencies of repeated events such as motor accidents. I once feared that it was impossible to infer relative probabilities from unique events, but I am now persuaded that it can be made mathematically respectable. Cosmologists have used similar arguments about the improbability that extraterrestrial intelligence exists. On Earth, it has taken a long time for a species to evolve that might be extraterrestrially detectable: a species with rocket launchers, radio telescopes and transmitters. It has taken a substantial fraction of the life of the solar system for us to show up, suggesting we are not an evolutionarily probable life form. A search for extraterrestrial intelligence may be technically easier than a search for extraterrestrial life, but it is a search for something much less likely to exist. The argument uses the time of human evolution, which is a unique event, to infer the chance that extraterrestrial intelligence will evolve.

We use analogous arguments in more familiar walks of life. If, for instance, you receive a 'chain-letter' offer, you can reason as follows. The number of recipients of the letter increases with time (to be exact, it increases exponentially). The early recipients will, if they take up the offer, receive money; but the late recipients would be taking up the offer just before the chain collapsed, and would lose money. When you receive the letter you are not told whether you are early or late in the chain. But you can infer, simply from the exponential increase with time, that you are more likely to be late than early in the chain. You bin the letter. There is a story about a wealthy capitalist – a Rockefeller, when I was told the story – who sold out just before the 1929 stock-market crash. He was having his shoes polished, and the boy gave him investment tips while doing his job. Our Rockefeller deduced that when even the shoeshine boys are in on the game, there cannot be any potential buyers left, and it is time to sell. He was using a unique observation to infer a probability. The inference is not exactly the same for a chain letter and for evolution, but there is a common basic logic. We can use the times of unique events such as the origin of life and of complex life to infer the probabilities that these events would occur. If someone thinks that life will, as a matter of course, or even probably, evolve to be complex, ask them why it took so long for life forms made up of anything more than one cell to appear on Earth.

The probabilities of evolutionary events can be imagined in two ways. I have mainly talked about them in terms of how long it takes for the event to happen: an improbable life form in this sense is one that took a long time to evolve. But, as the cosmological argument illustrates, we can also think in terms of the chance that the life form will evolve at all. If we imagine the evolution of life on many planets, what fraction of those planets are likely to have cellular life, complex multicellular life, life with blast-off technology? Improbable kinds of life will exist on only a few planets, probable kinds on many. Life will then exist all over the Universe, but complex life will be rare, and freaks like you and me who belong to a species that may be intergalactically detectable will be seriously rare.

I do not think that the arguments in this chapter conclusively show that the evolution of complex life is a paradox. Nor, I suspect, are they the only arguments that could be made. However, the unfolding discoveries of complex life since Antoni van Leeuwenhoek visited Berkelse Mere, the philosophical clarification since Darwin of the meaning of life, the shrinking evolution of molecules in elemental test-tube experiments, and the relative timing of the origin of life, and the origin of complex life, in the 4000 million years of evolutionary history, all suggest that complex life is something of a puzzle. Complex life is a problem to be explained, rather than an inevitability we can complacently observe and intellectually take for granted.

2
The gene number of the beast

The geneticists in the counting-house

Complexity is an ill-defined term, and I have been tempted to avoid it completely. I do not think it is meaningless to say that some forms of life are more complex than others, but I am as puzzled as anyone by what exactly I mean when I say it. There is no biologically agreed definition of complexity, but I suspect most people would agree on what should contribute to it. Structural complexity is one factor: the number of organs, tissues, or types of cell in a body. John Tyler Bonner once estimated the number of different types of cells in various creatures, as an index of complexity. Examples of cell types include bone cells, brain cells, blood cells and skin cells. He estimated that humans contain about 120 different types of cell. There is more than one way of defining what a cell type is, and other experts estimate that we have 200 or more cell types, but I shall stay with Bonner's number to make a consistent comparison. The big division among animals is between vertebrates and invertebrates. Vertebrates are the group of fish, amphibians, reptiles, birds and mammals, including ourselves; invertebrates are all the rest – the animals without backbones, such as shrimp, spiders, starfish, squids and worms. All vertebrates, in Bonner's estimate, have the same 120 or so cell types as us; they are all equally complex by this criterion. Most invertebrate animals have about fifty-five cell types, except for the simpler invertebrates such as jellyfish who have eleven or so cell types. Flowers have about thirty cell types, mushrooms seven, and bacteria one, or at most two. The number of types of cell therefore fits in with an intuitive idea of complexity.

The complexity of behaviour, particularly social behaviour, is another factor. It could be measured by the variety of behaviour patterns and social interactions that the members of a species act out. There is also the number of stages in the life cycle. Humans in a sense have only one life stage: we are born as mini-humans, and grow up. A butterfly has three: one from egg to caterpillar, then the chrysalis, then the imago. A life form with more stages is more complex than one with less, and by this criterion the world champions of live complexity are, as George Williams once argued, certain parasitic worms. The liver fluke is a 'lowly and simple' animal (in Williams's ironic phrase) that eventually grows up to occupy the liver of a sheep, but it has quite some experiences on the way. Here, in paraphrase, is Williams's description of its life cycle. It starts as a fertilized egg in the water of a pond or stream. It develops into a ball of cells, which swims by means of tiny oars, called cilia, on the outside of the ball. This ball of cells can detect snails in the water and will penetrate a victim if it can find one. There it morphs into a new, snail-infesting form, which buds off many copies of itself. These morph again, into a form that moves about inside the snail and reproduces by yet another procedure. These offspring eventually morph into a further form, which is designed to burrow out of the snail and swim to a blade of grass at the edge of the water. It swims by quite a different method from the oar-powered ball of cells; it moves by wriggling its tail. Once it is attached to its blade of grass it develops into a dormant stage, consisting of a nondescript heap of cells. These cells remain on the blade of grass until they are eaten by a sheep, when they travel to the sheep's liver and morph into a young fluke. The fluke is a worm-like creature, and grows up into an adult fluke. The cycle is complete. You can imagine what these protean flukes would think of some human who made absurd noises about humans being the pinnacle of living complexity – humans, who just produce little humans, generation in generation out, and have Kafkaesque vapours even at the thought of a more complicated life. If humans could produce fish as babies, and the fish could metamorphose into wheel-animalcules, one might pay attention to them, but really . . . their whole lives hardly amount to more than one school grade in a serious specimen of animal life.

Complexity might, therefore, be measured by including both instantaneous structural and behavioural complexity and the number of stages in the life cycle. An informal definition of complexity along these lines is how long it would take someone to describe the life form completely: to write out a description of its anatomy, life cycle and behaviour. A more formal version of this description-length criterion comes from information theory, and counts the number of 'bits' of information in a body. A 'bit' of information is the answer to one yes/no question. John Pringle defined the complexity of a living creature as 'the number of parameters needed to define it fully in space·and time'. This, and other informational definitions, were suggested back in the 1950s, as soon as information theory was applied to biology; but they have proved difficult to apply. A bacterial cell, for example, ought to be an easy case. One early estimate found 10^{12} bits of information in the arrangement of molecules in a bacterial cell, but this can be reduced to 10^4 bits if many arrangements of the molecules would all permit the bacterial cell to live: this small difference, amounting to eight orders of magnitude, suggests there may be some room for sharpening up the estimates. (The more familiar computer 'byte' equals 8 bits. The estimates for bacterial information correspond to around 125,000 megabytes and 1.25 kilobytes respectively.) The calculation of the number of bits in living creatures remains a problem for the future.

Cell types, developmental stages, information theory: these are all strong contenders for the definition of live complexity, or components of a definition. They are related, in that a body with more cell types would also contain more bits of information. In this book, however, I am not going to argue about their relative merits, but concentrate on what may be a deeper criterion of complexity. I shall be more or less identifying the complexity of a life form with the number of genes it contains. On this definition, a simple life form has fewer genes than a complex life form. Therefore, if while reading this book you ever find yourself thinking 'but what does complex *mean*?' remember that in all the arguments we shall meet 'gene number' can be substituted for 'complexity'. Gene numbers are closely related to the other criteria of complexity. Stuart Kauffman, of the Santa Fe Institute, has found an approximate mathematical relation between the number of genes and number of cell types in a range of life forms: the number of genes equals the square of the number of cell types. Kauffman's exact relation is questionable. If, for example, you try to confirm it with the numbers for cell types at the beginning of this chapter, and the numbers for genes that we shall come to in a minute, you will find it does not work. That is because Kauffman and I used different sources for the numbers: gene and cell-type numbers are not well enough known for us to be sure about them. Nevertheless, it makes sense that there should be some relation between cell types (and developmental stages) and gene number, because a body with more cell types needs more genes to code for that cellular information. There are some snags in the gene–number criterion, but we can use gene numbers in a fairly full, if not final, account of the evolution of live complexity.

A gene is a piece of code in the DNA. Genes can be thought of as rather like instructions. A gene that codes for a digestive enzyme in the stomach is like an instruction which says 'digest that sugar'. To be exact, the digestive enzyme is a particular protein and the gene contains the codes that describe how to assemble the protein. Gene codes are a series of letters, such as ... AAGCTGATA ... and each triplet of letters in the sequence (such as AAG or CTG) codes for a unit in the protein's structure. A protein is read off from a gene, and the protein then carries out some function in the body. Genes that work in the immune system code for proteins that do things like binding disease-causing parasites, presenting the bound parasites to executioner cells, and chemically dissolving the parasites. These activities correspond to instructions both direct and implicit like 'detect the enemy', 'here's one of the bastards', and 'kill'. The genes that work during development are more difficult to imagine, but they again work very like instructions. One set of genes is switched on early in development and sets out the basic map of the body. For instance, a gene may be switched on at one end of an early embryo, and produce a particular protein. The protein may then diffuse away from the region where it is produced, setting up a gradient. The protein has a high concentration at the end where it is produced, and a low concentration at the other end of the embryo. This gradient might specify the axis from head to tail. Another protein might make a gradient from back to front of the body. The genes that code for the gradient proteins are as much like signposts as commands, saying things like 'this side up', 'this way to the head', or 'this way to the toes'. Sub-instructions can then be expressed, depending on where they are in the embryo. A sub-instruction gene may be switched on or off, depending on the concentration of the gradient protein around its cell. The sub-instruction then says something like, 'if you are at the toe end, start to grow a digit; but if you are at the head end do nothing'. The 'grow a digit' instruction would presumably work by activating a whole set of sub-sub-instructions to make cells develop into bones, skin, toenails and so on. Gene numbers are related to complexity because, generally speaking, a more complex life form uses more instructions than a simpler life form.

Geneticists have measured the number of genes in the DNA of several creatures. Viruses use the least. The human immuno-deficiency virus has seven genes; the influenza virus has eight. They have only minimal instructions, saying things like, 'take me in', 'copy me', 'make a protective wrapper' and (possibly) 'sneeze'. Bacteria are tiny cells but have many more genes than viruses. The bacterium that causes syphilis has about 1000 genes, the famous *E. coli* has about 4300. (*Escherichia coli*, or *E. coli* for short, may be biologically the best-understood life form on this planet. It is normally a benign inhabitant of our lower guts, but dyspeptic, even lethally dyspeptic, strains do exist, such as strain 157, and *E. coli* has an unfortunate association with media food scares, at least in Britain.) *Escherichia coli* may be towards the high end of bacterial gene numbers, and something like 2500 may be a more typical gene number for bacteria. A bacterium is a whole self-contained cell and needs all the instructions to build, fuel, maintain, repair and defend a cell, as well as to reproduce. The next most complex cell that has had its genes counted is yeast. There are actually many kinds of yeast, but the standard laboratory yeast is bakers' (and brewers') yeast. It is single-celled, but the cell is eukaryotic, not prokaryotic. As we saw in Chapter 1, eukaryotic cells are larger and more complicated than prokaryotic bacterial cells, and it takes more instructions to code for the extra complexity. Yeast can respire aerobically, using its mitochondria, when oxygen is present; it is then in bakers' yeast mode. Or it can respire anaerobically, when oxygen is absent; it then produces alcohol, up to about 13% depending on how much sugar is present, and is in brewers' yeast mode. Yeast contains about 6000 genes.

Ciliates are also single-celled eukaryotes, but larger and more complex than yeast; they are about the simplest animalcules van Leeuwenhoek saw. Ciliates use 12,000–15,000 genes. These numbers are about the same as those for animals without backbones. Fruit flies use about 14,000 genes and worms about 19,000. If the similarity of gene numbers in single-celled and many-celled invertebrate eukaryotes is confirmed by future research, it will be a striking generalization. It will imply that it takes hardly any extra genes to control the development of a multicellular animal on top of the genes needed to control a cell. I should have guessed otherwise. Flowering plants have a basic set of about 20,000–25,000 genes, similar to the number for invertebrate animals. Vertebrates seem to have more genes still. The estimates for gene numbers in the puffer fish, mouse and human are all similar, in the 50,000 to 100,000 range. The number of genes in a human is usually put at between 60,000 and 100,000, but we shall not know the exact number until the human genome project is complete. I suspect our gene numbers will turn out to be at the lower end of the range, and in this book I shall use 60,000 when I need a figure for the number of genes in a human being.

Gene numbers broadly conform to our intuitive ideas of complexity. The pattern of the numbers for bacteria, yeast, worms and humans looks about right. But some of the subpatterns are more surprising. The similarity between ciliates and flies is one oddity, and not all biologists would expect the increase from flies to fish. The insides of a fish are not clearly more complex than those of a fly (though fish have more cell types), and insects usually have more complex life cycles. There may be more to complexity than gene numbers. Maybe we also need to know about the complexity of the genes themselves, and a gene in a fruit fly may code for a more complex instruction than a gene in a ciliate. Or maybe we need to know how the genes interact, and a liver fluke or a butterfly can squeeze more complexity out of its genes than a fish does. Or some of the genes in a fish, or in a ciliate, may be redundant or trivial. The number of genes that matters (and that is meant to be expressed in the estimates we just looked at) is the number of different genes. If a library contains a thousand copies of one book (for example, *The Collected Speeches of Comrade Stalin*), there is a sense in which it has only one book – and this sense is analogous to the way we are counting genes here. Humans contain two copies of every gene, one from their father and one from their mother; we therefore divide the total number of genes in a person by two. There are further corrections before we arrive at the final gene number of 60,000. For example, humans inherit multiple copies of some genes from both parents, and again the multiple copies need to be counted once each. The details do not matter, just the point that 'number of genes' means 'number of different genes'. This is clear enough in theory but correction can be difficult in practice. If the multiple copies of a gene are practically identical, they can be recognized like multiple copies of Stalin's speeches and counted as one. But they may be redundant in less obvious ways, as if one were a volume of Stalin's speeches and another were a volume on the same topic by one of

33

Stalin's cronies. The gene numbers of ciliates could therefore turn out to contain more concealed redundancy than those of the fly. (Ignorance cuts both ways. Flies may turn out to contain more genetic redundancy than ciliates, and the puzzle will be compounded.) For all these reasons, the relation between gene numbers and complexity is not completely understood. However, I do not need to answer all the questions here – much as I should like to. The exact gene numbers are mainly illustrative and are not crucial in the arguments that follow. Gene numbers are already a good working definition of live complexity. They may turn out, after another ten years of genomic research, to be close to a perfect definition; but we do not need to anticipate any such outcome here.

Accidents at the gene-copying machine

If human beings and other vertebrates have twenty to fifty times as many genes as a bacterium, gene numbers must have increased during evolution. How does this happen? In general, evolutionary change has two unrelated components. The first is variation – inherited variation – within a species. For the evolution of gene number, some individuals in a species must have one quantity of genes (such as 1000) and other individuals other quantities (such as 950, 980, 1020, 1050); and the offspring of the 950-gene individuals must on average have fewer genes than the offspring of the 1050-gene individuals. The second component is natural selection: some kinds of individual have to leave more offspring than others. Gene numbers will evolve up if the individuals with more genes than average leave more offspring than average, and down if those with fewer genes leave more offspring. Either way, the species is evolving. We therefore need two kinds of theory, one about how gene numbers can vary among individuals to begin with, and another about why natural selection should favour one number of genes rather than another.

Individuals with new numbers of genes result from copying accidents (called mutations) during reproduction. The DNA is meant to be copied once during reproduction, but accidents do happen. Mutational accidents that increase the number of genes are known as duplications, and happen when all or part of an organism's DNA is copied twice. One kind of duplicative accident is like a common kind of mistake made by human scribes and typists. A page of written text can by chance have the same word, or row of words, in the same place on two consecutive lines. It is then easy to jump down, missing out a line of text, or up, to make two copies of the line between. Imagine a scribe with his quill, which he has plucked from the monastic cock (a distinguished bird, with a pedigree tracing back to the greatest of cocks – the one who caused St Peter to weep bitterly). He is copying out the Ten Commandments:

> Thou shalt not commit adultery.
> Thou shalt not steal.
> Thou shalt not bear false witness against thy neighbour.
> Thou shalt not covet thy neighbour's house, thou shalt not covet
> thy neighbour's wife, nor...

His concentration lapses for a moment in the 'Thou shalt not' region and he jumps down one line. Life will be 10 per cent simpler for his readers. Or he may jump up a line, and copy a commandment twice. In human copying, repeats are easier to spot, and so fix, than are deletions, and our Ten Commandments are more likely to have been reduced from an originally more comprehensive list than expanded from a shorter list originally dictated by an anarchist divinity on Mount Sinai. Indeed, a reduction of the list by scribal error would explain the otherwise baffling omission of several antisocial habits. An expansion is more likely if the scribe makes two mistakes, one that copies out a line twice and another that introduces perhaps a small spelling mistake into one of the duplicated words. The proliferation of commandments to do with neighbours at the bottom of the list is therefore suspicious. I wonder whether the Hebrew words for 'house', 'wife' and 'bear false witness against' are at all similar.

In a stretch of DNA code, a similar series of letters may be repeated, such as in TACCGC ... [many letters] ... TACCGC .... The DNA copying machinery is liable to jump from one to the next, deleting a bit of code, or to jump back and reduplicate the code in between. The exact mechanical reason why these mistakes happen is different from the visual confusion in a monk's erring eye, but the accidents are closely analogous. The passage between two similar bits is liable to be duplicated by the DNA copying machinery, just as by a human scribe. In DNA copying, unlike human copying, repeats are about as likely to happen and

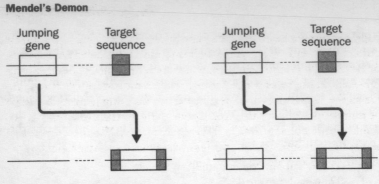

Figure 2. Jumping genes may copy themselves (*a*) conservatively, without duplication, or (*b*) multiplicatively. Multiplicative jumping usually works via an RNA intermediate.

be let through as are deletions. And whereas a scribe's errors are most likely over the region of a line or so, any length of DNA can be duplicated or deleted. Short stretches of DNA may be missed, or copied twice. So may a whole gene. If the members of a population at one time have 10,000 genes, a mutant individual might be produced in which one of the genes had been accidentally copied twice; the mutant would have 10,001 genes. A region of the DNA might be accidentally duplicated; if the region contained 1000 genes the mutant individual would have 11,000 genes. On the largest scale, all the DNA might be duplicated, creating a mutant individual with 20,000 genes – all the 10,000 of everyone else in the population, but in two copies. These kinds of duplication have all been shown to occur, and at rates that are high enough to explain evolutionary changes in gene numbers.

Genes can be duplicated by rogue copying, as well as by the innocent copying accidents we have discussed so far. The DNA of most species of complex life contains many peculiar 'jumping genes'. Jumping genes are sequences of DNA that have the power to copy themselves elsewhere in the DNA of one organism. They are probably molecular parasites, which persist – even proliferate – only because they have not been removed by the host. Some jumping genes move conservatively: they excise themselves from one site and jump to another (Figure 2). They do not matter here. Others do so multiplicatively: they make one or more copies of themselves without excising from the initial site, and then the copies insert themselves elsewhere in the DNA. This is a kind of duplication. After it happens the individual has a duplicated stretch of DNA. The jumping gene may also pick up a bit of neighbouring DNA and take it for a ride, and then that bit is duplicated too. As far as the whole organism is concerned, jumping genes are a kind of mutation and we can think about them in the same way as we think about conventional copying accidents. We therefore have several processes – more and less accidental – that produce individuals with new numbers of genes.

Modernizing Moses

When a duplicative mutation happens, it merely produces an individual with extra DNA. Mutations alone are not enough to explain evolution, in gene number or in anything else: for evolution to happen, the mutation has to spread through the population. It will do so if the mutant individuals leave more offspring on average than the other members of the population. Why should this be? We can start with mutants in one, or a small number of, duplicated genes. The mutant individual with two copies of a gene is not obviously going to be more successful than average. He or she contains duplicated instructions, like a Bible that commands 'thou shalt not commit adultery' twice rather than once. At best the duplication is harmless, at worst it is absurd. Natural selection probably usually removes the mutations that have two copies of a gene. Nevertheless it can also sometimes favour them. More of the gene's product may be needed, as if people actually need to be told twice not to commit adultery before they get the message. The gene may be needed in two circumstances, and the mutant duplicate gene is able to work in the second circumstance. For instance, you may be telling people from the pulpit not to commit adultery. It seems that they are behaving themselves outside the workplace, but then forget the instruction when they go to work. You need to repeat it there too, perhaps in compulsory sex education classes. The duplicated instruction is advantageous by virtue of its repetition.

Another possibility is that the duplication creates something new and useful, by luck. Fruit flies have an interesting gene called *jingwei*. The middle of *jingwei* looks like another gene, well known to geneticists, that codes for a substance called alcohol dehydrogenase. Alcohol dehydrogenase has a sobering effect in humans, and probably in fruit flies too. Fruit flies need something to keep them sober, because they live around vineyards and also gatecrash alfresco parties in the summer, provided there is wine. In any case, *jingwei* originated when the alcohol dehydrogenase gene duplicated itself into the middle of an unrelated gene called *yande*. The hybrid creation probably did something new and different from either alcohol dehydrogenase or *yande*. The exact nature of the new skill is uncertain, but it was beneficial somehow and natural selection established it in the fruit fly species.

Biologists like the idea that jumping genes could create advantageous new genes in this way. The idea is attractive because when the DNA of two very different species (such as worm and yeast) is compared, it looks as though many of the extra genes in the worm contain bits of yeast genes that have been mixed up. The DNA of the worm contains not only whole genes that look like genes in yeast, but also bits of yeast genes that have jumped into a new background. It is as if the worm's DNA contained not only the Mosaic decalogue but also these splicings of Clough:

> Thou shalt not kill; but need'st not strive
>> Officiously to keep alive.
> Do not adultery commit;
>> Advantage rarely comes of it.
> Thou shalt not steal; an empty feat,
>> When it's so lucrative to cheat.

and of Auden:

> Thou shalt not do as the dean pleases
> Thou shalt not write thy doctor's thesis...
>> Nor with compliance
> Take any test. Thou shalt not sit
> With statisticians nor commit
>> A social science.

I do not know how natural selection would work on Clough's and Auden's instructions, but the underlying process of jumping genes in principle can invent instructions that are useful as well as new. The scrambled genes that we now see in the worm could have arisen, during evolution, if jumping genes had duplicated bits of genes and scattered them around the DNA. Jumping genes then have creative long-term effects for the species that they occupy.

So far I have been talking about duplicated genes, or parts of genes. In other circumstances, it may help to duplicate all the DNA. An example is when members of two different species interbreed and produce sterile hybrid offspring. In human beings, species membership is clear-cut: any human can interbreed with any other human, subject to obvious provisos – they have to be of opposite sex and of reproductive age. But humans do not interbreed with members of the nearest related species, chimpanzees. Humans are one species, chimps are another. In other forms of life, however, species seem to be less clearly divided: an individual usually breeds with a member of its own species, but may breed with a member of a closely related species. There is a problem about what 'species' refers to here, but we need not enter into it. When different species – or perhaps they should be called near-species – interbreed, the hybrid offspring may be OK; or they may be inviable; or (like mules) they may be viable enough, but sterile. If the hybrid offspring are all OK, there is no advantage to doubling the DNA; but there can be an advantage in some interspecies hybrids that are viable but sterile (though it does not help in mules).

The reason is probably that doubling the DNA solves a problem in chromosome pairing. An individual's genes are arranged in a number of chromosomes. Human beings, for example, have 46 chromosomes: two sets of 23. A chromosome is rather like a book of instructions, and if we have 60,000 genes in our set of 23 chromosomes, one chromosome carries an average of about 5000 instructions. The double set of chromosomes is reduced to a single set during reproduction. A hybrid formed between a human and a near-human would contain 23 chromosomes from the human parent and 23 from the near-human parent. The hybrid has a full gene set for making a human (or a near-human) and may be viable enough. But what will happen when it comes to reproduce, or try to reproduce? It has 46 chromosomes, and has to reduce them to a 23-chromosome set in its eggs or sperm. The 23 that are needed, however, are not any old 23 out of the 46, but the 23 that make up a complete gene set.

Suppose you have two sets of a 23-volume encyclopaedia and you are going to give away a spare set: you need to give away the

23 volumes that make up a complete encyclopaedia (and leave you with a complete encyclopaedia). You do not want to give away 23 randomly picked volumes from your 46 – both copies of volume 1, one copy of volume 2, no copy of volume 3, and so on. In biological reproduction, the genetic machinery achieves the correct division by what W. D. Hamilton, in a memorable image, called 'this gavotte of chromosomes'. There is a special cell division, called meiosis, in which the chromosomes move about in an elaborate and inflexible manner. The chromosomes first pair up (volume 1 arm in arm with volume 1, volume 2 with volume 2, and so on). A set of cables is constructed inside the cell. The chromosome pairs then neatly align at the cell equator and are attached to the cables by a special, motile molecular engine. The engine rotates, the chromosomes move apart, and various surveillance procedures ensure that the right chromosomes end up in the right daughter cells. The result is two perfect chromosomal sets at either end of the cell, and the cell can then divide. The pairing up into two full sets is the key to the correct division into two complete chromosomal sets; if a chromosome paired with the wrong partner, the whole division would mess up. But how are the members of a pair recognized? It seems to be a matter, at least in part, of similarity between the chromosomes, or of a region within them. The two chromosomes form a pair provided that they are more similar to each other than to any other chromosome – as they usually are, just as (in the encyclopaedia analogy) two copies of volume 1 are more similar to each other than either is to any other volume in the two sets. The gavotte then goes off well. But in a hybrid the two chromosomes of a pair may differ. Hybrid offspring contain chromosomes from two parental species. The copy of chromosome 6 from one species may be more like the chromosome 10 from its own species than the copy of chromosome 6 from the other species; it will then pair with the wrong partner. Instead of two complementary rows on the gametic dance-floor, we end up with two higgledy-piggledy sets. On one side, a volume 6 is clasped in a horrid embrace with a volume 10, another odd volume 6 is dancing by itself, and (perhaps) volume 3 is missing. On the other side there are two copies of volume 3 and no volume 6. The offspring will inherit a mismatched set of encyclopaedia volumes. In biological terms, the offspring will lack some genes, or have excess copies of others. The offspring may be sick, or

never be born. If the hybrids' offspring are never born, those hybrids are sterile.

Many hybrids are sterile. They are lost to evolution and never seen again. But fertility can sometimes be restored by doubling up the DNA in the hybrid. The hybrid then has four sets of chromosomes, two sets from one parental species and two sets from another parental species. The task in reproduction is to reduce the four sets to two double sets. The chromosome 1 from one parental species no longer has to make the difficult pairing with the insufficiently similar chromosome from the other parental species. It can pair with the other identical copy of itself. There will now be 46 instead of 23 chromosome pairs lined up at the equator of the cell, each pair made up of two identical chromosomes. When the pairs divide, the result is two perfect chromosome sets. Natural selection can therefore favour a wholesale DNA duplication after two species have hybridized. Plants have particularly gone in for this process; animals less so. Many garden flowers and agricultural crops originated as hybrids; they were made by crossing two different parental species and multiplying up the chromosomes.

Mergers between species are another mechanism by which gene numbers can change. Mergers particularly matter for us because (as we saw in Chapter 1) they probably contributed to the origin of the eukaryotic cell. As with gene duplications, we have two questions: how did the new type of individual with extra genes come into existence, and why did natural selection cause it to spread? There is an interesting conceptual difference from the gene duplications. Gene duplications arise by a genetic accident, like the scribe who copied a commandment twice by mistake. In a symbiotic merger, the accident that alters the gene numbers is more ecological. An engulfed prey bacterium might have managed to survive inside a predatory proto-eukaryotic cell, perhaps following a period of evolutionary change in the bacterial species after its members had suffered repeated predatory attacks. The total number of genes in the cell-containing-a-cell would be increased: it has all the genes of the predatory cell plus all the genes of the undigested prey. The ecological act in which one cell engulfs another and then suffers indigestion would be conceptually analogous to the genetic accident in which a gene is copied twice. Why should the merged cell be favoured by natural selection? The answer is probably that the two partners had complementary physiological skills. One partner, for instance, might throw out sugars as a waste product, whereas the other might be able to burn the sugars as fuel. The merger would be synergistic, and natural selection would favour it. But whatever the exact advantage was for the merged cell, the merger itself increased the number of genes, and mergers in general provide another mechanism by which gene numbers can change.

From 0 to 60 in 10^{17} seconds

The historian of live complexity would like to know when, during evolution, gene numbers have increased. He or she will also want to know whether the increases have been by many duplications of scattered, small regions of DNA, or by a smaller number of wholesale duplications of the whole DNA, or by symbiotic mergers. The history of gene duplication is an emerging 'hot topic' in biological research, and so far we only know some sketchy features of the history, not the full story. There has probably been only one big merger event in the history of complexity, the merger that created the eukaryotic cell. Gene numbers were certainly increased by this event, but we do not know by exactly how much. The gene numbers we looked at earlier give us some idea. Bacteria, we saw, have about 2500 genes each. A merger between two bacteria could create a cell with 5000 genes, and it is noteworthy that the simplest eukaryotes such as yeast contain about 6000 genes. In the grand sweep of history, from the origin of life when there may have been one gene or less, up to modern humans with 60,000 genes, mergers may have contributed only 2500 or so of the genes – less than 5 per cent of the complexity increase.

Most of the remaining increases have been by duplications of various fractions of the DNA within a life form. Jumping genes, duplicated genes and duplications of whole DNA sets have all contributed, but the full story remains to be worked out. It used to be thought, for instance, that the four-fold increase in gene numbers between invertebrate animals, such as worms and flies, and vertebrate animals, such as fish and us, was caused by two rapid rounds of doubling of the whole DNA set, at some time near to the point when the vertebrates first evolved. But careful inspection of vertebrate DNA now suggests that the increase was produced by a larger number of smaller duplications.

Biblical scholars have been able to reconstruct the authorial history of the Bible, from a close reading of the text itself and a knowledge of external historical events. The book of Genesis, for instance, contains two contradictory accounts of the creation in Chapters 1 and 2. It is reasonably certain that the two accounts are derived from a merger event rather than a duplication. The merger event can be traced through the whole book of Genesis, not just the first two chapters; it probably happened when a writer, some time before 400 BC, combined two other authors' work into one book. As our knowledge of DNA sequences improves, we should be able to read the human genome like textual scholars. We should be able to discern the mergers, duplicating accidents, shufflings and transpositions by which modern human DNA has evolved since the origin of life.

Explaining complexity

When natural selection establishes a duplicating accident in all members of a species, perhaps because it restores a barren hybrid to fertility, the event does not increase live complexity, nor happen in order to increase it. The organisms with their doubled-up DNA are no more complex than before. A Bible that says 'thou shalt not steal thou shalt not steal' is more repetitive, not more complex, than one that says it once. The mere multiplication of some DNA does not create complexity. It matters only because it may lead to the creation of complexity. Once an extra set of genes has been established in a species, it may provide the raw material for a future increase in complexity, if natural selection favours an increase. We need a further theory, one that explains why the extra genes should be put to use in the evolution of a more complex creature.

A full explanation of an evolutionary change in complexity has three parts. We have dealt with two of them: the explanations of gene number mutations within an individual (by jumping genes, accidental duplicate copies, and mergers) and of why natural selection favours increases in gene numbers (by symbiotic synergy, amplified gene product, creative gene combinations, and hybrids that can breed). The third part is an explanation of why the extra genes should evolve to code for a more complex life form. The extra genes did not evolve for their pleasing future consequences, such as to create the complex life we now find interesting. Evolution, in Darwin's theory, is short-termist; it works opportunistically, with piecemeal changes and immediate advantages, not with long-term strategic plans.

What, exactly, is the kind of event that the third theory is going to explain? Consider a gene duplication. The duplication itself evolved to fix some more immediate problem, such as a chromosomal snarl-up in an interspecies hybrid. The rescued fertile hybrid is no more complex than either of the species it evolved from. But the hybrids with duplicated DNA have two sets of genes where in a sense they only need one set. After a period of evolutionary adjustment, the second set may no longer be needed for the hybrids to breed. The chromosomes may evolve in such a way that a single set of chromosomes can both pair up properly and code for a full, fertile organism. There are now loads of excess genes in the creature, and these genes can evolve in either of two ways. They may be lost over time, and the complexity of the life form will then remain unchanged. Alternatively, some of the extra genes may evolve a new function, and complexity will increase. The evolution of a new gene function is a distinct process, requiring some reason for natural selection to act after the duplication has been established. We need the third kind of theory to spell out what that reason is. The duplication sets the stage, but the play still has to be written.

The relation between the evolution of complexity and gene numbers is simpler for a symbiotic merger. In this case the extra genes already have a distinct function, unlike the multiplied identical genes after some DNA is duplicated. In the merger that led to the eukaryotic cell, there would initially have been a rare 'mutant' cell-within-a-cell form in a population of simpler cells. This was formed by an ecological accident, when one cell ate but did not digest another. Natural selection then favoured the merged form, probably because of some synergy in its merged metabolic set-up. The metabolism of the merged cell was already more complex than that of either partner alone. In this case the second and third theory are one and the same, but this does not contradict the point that natural selection lacks future sight: the increased complexity here has an immediate advantage as the gene numbers increase. Moreover, the first and second theories are still separate: the big bacterium ate the small bacterium to obtain its dinner, not to obtain codes that might make itself more complex.

It is not enough to point to the basic mechanisms of evolution – mechanisms such as mutation and the short-term workings of natural selection – to explain complexity. We also need a theory about why natural selection should favour a more, rather than a less or an equally, complex form. We are almost in a conceptually unique position in the history of human thought. It is an old idea that life is arranged in a great chain of being, or *scala naturae*, from the simplest, lowest form of life up to higher and more complex creatures like us. Inspired thinkers saw a further chain of angelic beings that connect us to God. Arthur O. Lovejoy, in his book *The Great Chain of Being*, traced the idea back to ancient Greece and described it as 'one of the half-dozen most potent and persistent presuppositions in Western thought'. Darwin's theory logically destroyed the idea, as Darwin well knew: 'it is absurd,' he said, 'to talk of one animal being higher than another.' But the idea was stronger than Darwin's logic in the late nineteenth and early twentieth centuries. Evolution came to be imagined as a progressive ascent, for reasons not of natural selection, but some grand historic or divine plan. The evolution of a species was then like the development of an organism from egg to adult, and it was no more remarkable that later evolutionary forms were more complex than that an adult is more complex than an egg. Some force immanent in life caused the simple (or lower) life forms to evolve into something higher and more complicated. The force was something like what we now call directed mutation: offspring tended to differ from their parents in the direction of the grand trend towards increased complexity. This force is completely different from Darwinian natural selection, but still it was highly popular and often confused with Darwin's own theory.

Quite why Darwin should have been hijacked in this way is an interesting historical question. Herbert Spencer is partly to blame. Spencer was the leading nineteenth-century exponent of the idea that evolution is inherently progressive, leading predictably to increasingly complex forms. He was also undoubtedly influential. (The historian Peter Bowler has even said that the Darwinian revolution should historically be called the Spencerian revolution.) But we should beware of shooting the messenger. Spencer's message was something people wanted to hear, and it enabled them to graft Darwin's theory on to the old stock of the 'great chain of being'.

45

Since that time, it has been repeatedly explained that Darwin's theory does not propose an ascending hierarchy from bacteria, through amoeba, worms and fish, to humans. No serious thinker believes this any more, and what would, for most of human history, have been the obvious solution to the problem of complexity is no longer available to us. It is no use looking for a solution in the basic forces of evolution; they have no inherent tendency to generate complexity.

I can make the same point in the ersatz philosophical language of reductionism. We have three kinds of theory. One is about how gene numbers change within an individual; it is a theory of mutation. A second is about how gene numbers change over evolutionary time in a species. A third kind of theory is about why species evolve to become more complex, given that they have the genes that make it possible. In most non-Darwinian explanations of complexity, the first kind of theory is enough: the mutations alone drive the evolution of complexity. The evolution of complexity can be reduced to the mutation process. In the Darwinian explanation of complexity, however, the third kind of theory does not reduce to the second, and the second does not reduce to the first. We have to work out all three on their own merits.

The pioneering spirit

An increase in gene numbers provides the raw material for the evolution of complexity, but what determines whether or not it will be used? Here is how I think about the question – as indeed does almost everyone else who thinks about it nowadays. A species at a time will contain a certain array of forms, which exploit a certain range of resources. This range of resources is what ecologists call the species' niche. In birds, for instance, the form of the beak influences the food that the bird can eat. This has been shown in many species, including tits in English woodlands and finches on the Galápagos Islands. Birds with smaller beaks eat small food, such as seeds, more efficiently; birds with larger beaks eat larger food more efficiently. Within one species the range of beak sizes determines the range of resources that are consumed. Beyond the beak size range of one species, at both ends of the range, there will be further species – a smaller-beaked species, eating yet smaller seeds, at one end, and a larger-beaked species at the other. If the food available in the environment is constant, the birds' beaks will probably also stay constant. But evolutionary opportunities will arise from time to time. The food supply might expand at the border between two existing bird species – perhaps because the plants that produce seeds of the intermediate size are on the up. Then beak size might evolve up in the smaller species, or down in the larger species, to exploit the extra resources. Alternatively, one of the bird species might be struck by plague and go extinct; the other species could then evolve to occupy the newly vacated resources. In the case of birds, evolutionary opportunities are probably equally likely to arise on either side, above or below a species' existing range. It is as likely that resources will become available in one direction as the other. If we consider the future evolution of a species, beak size is about as likely to evolve up or down.

I have been talking about beak size only as a concrete example of what is a general idea. A similar story could be told for almost any attribute of a species, including complexity. Size is actually not unrelated to complexity, and we could conclude from the argument above that if a species evolves a new size it is about as likely to be bigger as it is to be smaller. Or we could rephrase the argument for something even more closely related to complexity, such as the number of life stages. The parasitic flukes we met earlier need to move from a snail to a sheep; the sheep incidentally ingest the parasites while eating grass. Now imagine that in a few million years' time sheep have changed during evolution and they no longer eat grass. We can imagine two possibilities: either that the future sheep eat snails, or that they eat rabbits. Stranger things have happened! In the world of snail-eating sheep, selection will force the parasites to lose two life stages, the snail-exiting and the grass-dwelling stages, decreasing their complexity. But the rabbit-eating sheep will create the opposite selective force (I am assuming that rabbits may sometimes take in the flukes from the grass). The flukes cannot enter the sheep via the grass, because sheep do not eat grass any more; but they may be able to get there via the rabbits that the sheep do eat. The flukes should now *add* a life stage, a rabbit-occupying stage, between the grass-dwelling stage and the sheep-occupying stage; they will evolve to be more complex. An environmental change that favours an increase in the number of life stages is about as likely as one favouring a decrease;

sheep are about as likely (or unlikely) to evolve to eat rabbits as snails. Complexity is equally likely to evolve up or down. I do not know that all evolutionary change can be understood in ecological terms, like the examples of beak size and fluke life cycles; but the idea is a general one. Changes in complexity then depend on changes in the environment that the species lives in. The forces that direct environmental change will often be unrelated to the complexity of consumer species, and future evolutionary changes in complexity should therefore be random: complexity may evolve up, down, or sideways, and not necessarily in that order.

We have a theory that suggests that evolutionary changes in complexity will be random. This may appear to contradict the observable fact that the average complexity of life has increased since its origin 4000 million years ago. But it is not difficult to reconcile random change within lineages and the grand historical pattern of live complexity. Life was originally simple. At the origin of life in the incipiently biotic soup, life was so simple that it bordered on chemical un-life. From there the only alternatives were for the complexity of life to increase, or to stay the same because it could not become any simpler. If it became any simpler it would drop back from life into chemistry. Simple forms of life are always with us. Viruses only a few genes long are with us and doing fine. Bacteria have dominated life on Earth for 3500 million years, and (as we saw in Chapter 1) there are fossil bacteria over 1000 million years old that are identical to modern bacteria. But in addition to the persistent simple forms, new and increasingly complex forms have been added over time. The reason why the average complexity of life has increased over time is that the range has increased while the lower limit has been fixed. We therefore need a small modification to the ecological theory that predicted random changes in complexity. At the very lowest limits of complexity, changes are more likely to be up than down; they are not random. Once life has evolved away from that lower limit, changes become equally likely to be up as down. That may be all there is to the historical rise of complexity. The idea certainly fits the historical facts: the range of live complexity has indeed steadily increased over time, over a fixed lower limit. It also fits with what we understand of the evolutionary process: changes are driven by ecological opportunities, and these will usually be equally likely to arise in either direction from a species' current state.

An additional force may operate that tends to favour increases, rather than decreases, in complexity in the species at the top end of the complexity range. An opportunity may exist to make a living beyond the range of existing species, because no competitors are established out there. If a species can evolve to be more complex than any existing species, it may find unoccupied ecological space to live in. New forms would then tend to be added during evolution at the top of the range. Bonner called this idea 'pioneering'. Pioneering drives species to occupy all the extremes of life, including the extremes of complexity.

Darwin seems to have thought along similar lines. His wording is a little rough, even cryptic; he was writing in a private notebook. Complexity, he noticed, tends to feed on itself, because the evolution of complex forms opens up new ways of living for yet more complex forms. I am not sure which (if any) of the ideas I have discussed here he was thinking about. He may have been thinking of some more general tendency of species to increase in complexity, not just species near the top of the range. But he was using much the same ecological framework as we have here. He wrote:

as the forms became complicated, they opened *fresh* means of adding to their complexity.   but yet there is no NECESSARY tendency in the simple animals to become complicated, although all perhaps will have done so from the new relations caused by the advancing complexity of others. It may be said, why should there not be at any time as many species tending to dis-development (some probably always have done so, as the simplest fish &), my answer is because, if we begin with the simplest forms & suppose them to have changed, these very changes tend to give rise to others.

The trend is towards complexity because the simple ways of life are all occupied. 'I doubt not if the simplest animals could be destroyed, the more highly organized ones would soon be disorganized to fill their places.'

I am distinguishing two theories about why complexity has increased during evolution (Figure 3). Bonner's 'pioneering' theory is slightly different from, but compatible with, what I'll call the 'random' theory. In both theories, most species in the centre of the complexity range are equally likely to evolve up or down in complexity. Also, in both theories, the species at the bottom of the complexity range cannot evolve to be any simpler. Bonner's theory

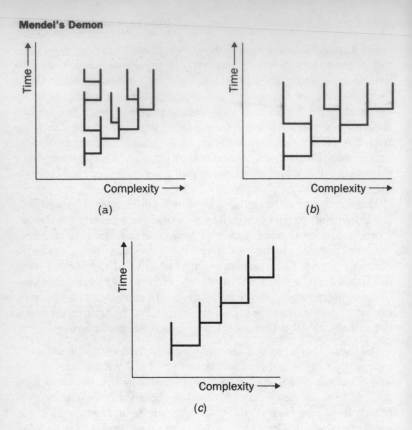

Figure 3. Three theories of how complexity evolves. In all three graphs the average complexity of life increases with the passage of time. (*a*) Random changes, but with a fixed lower limit. (*b*) Random changes in species of intermediate complexity; a fixed lower limit; and a tendency for more complex forms to evolve greater complexity. This is the theory of Darwin and Bonner. (*c*) Complexity tends to increase within each species. This could be for two reasons: (i) the discredited old idea that evolution is inherently directed towards increasing complexity; (ii) over a shorter period, in one group of life forms, natural selection might favour an increase in complexity. In all three, I have drawn an angular pattern of change, but that is not the issue: each right-angled line could be replaced by a diagonal line and the point of comparison between (*a*), (*b*) and (*c*) would be identical.

adds a suggestion about the species at the top of the range. They may be more likely to increase, than decrease, in complexity. The species that are already near the top of the range are likely to be the ones that do the pioneering because evolution usually proceeds in small steps: we, or some other great ape, are more likely than a worm or jellyfish to give rise to a more complex species than us. The history of complexity might be due either to random change plus a fixed lower limit, or to random change, a fixed lower limit and directional pioneering into unoccupied space at the top of the complexity range.

Which of the two theories is true? The answer depends on where the ecological opportunities lie. Andrew Carnegie once remarked, 'there ain't no money in pioneering.' No doubt capitalists do sometimes get carried away by business opportunities that appear to be completely new, particularly if they use new technology. In reality, it may be easier to make a living, or a profit, by exploiting existing opportunities better than the competitors rather than by seeking out unoccupied space. At least you know that a niche for a business does exist within the range of established businesses; the question is whether you can compete. On the other hand, you know that beyond the range of existing businesses you will not have to worry about competitors; the question is more whether a business niche even exists. Another expression was also current in Carnegie's time: 'Go West, young man!'

I doubt whether there is an advantage to pioneering in the bird-beak example. Any one species had a neighbouring species on either side, but this cannot be true indefinitely: eventually we must reach a species with the top-of-the-range beak. The reason no birds have bigger beaks is probably that there are not enough large food items to support them. The upper limit is then fixed by the resources available and no ecological opportunity exists for a species with a larger beak. But attributes other than beaks may allow more open-ended possibilities. Physical competitive power is a crude example. Imagine a set of species in which the individuals differ in strength. Members of the strongest species can beat up members of all the others, displacing them from desirable resources such as living space or water supplies; the second strongest species can displace all but the strongest, and so on down the ranks. It could then always be advantageous to be slightly stronger than the currently strongest life form, and the species near the top of the strength range would be more likely to evolve up than down in strength. It would not be advantageous to be *much* stronger than any existing form, because strength is costly to acquire and to maintain.

Now we have two stories, about beaks and about trials of strength: in the beak-size story the chance that a species will evolve up equals the chance that it will evolve down for all species; in the strength story, the same will be true for most species, but the species near the top may be more likely to evolve up. The question is which story has the greater generality, particularly for things related to complexity. The answer depends on what limits complexity at the top of the range. If complexity is limited by resource availability, as with seed sizes, evolution will not favour the pioneers. If the limit is only how complex life has so far evolved to be, because there is no advantage in being much more complex than anything else, then pioneering is more likely to be at work.

The case for 'pioneering' (Figure 3b) has to argue that evolution will more often be like the example of physical strength. The example is more general than it may sound. Real species rarely compete in such a crude way, but the argument works much the same with realistic kinds of competition. Human competitive power is based on intelligence, and a smarter life form than us could bully us as easily as we do chimpanzees (even though a chimpanzee is physically much stronger than a human being). As prey species evolve increasingly complex defences against predators, they create an opportunity for more complex predators: maybe the prey evolve socially cooperative defences and the predators then evolve socially cooperative hunting methods. Here is a further version of the argument. Many species compete ecologically by driving down resource levels, in much the same (if unintentional) way as businesses compete on price; only the most efficient operator will survive when resource levels have been driven lower than any other species can survive on. Competition of this kind may also sometimes favour evolutionary increases in complexity. Let's return to numbers of life stages. Suppose the members of an existing species of fluke have two stages, which live in two species of host. Some quantity of the two host species will be needed to support a viable fluke population, but these hosts may be in short supply. Now a mutant fluke acquires a third life stage, occupying a third kind of host. It can survive more easily because it can feed on a kind of host that is unavailable to the others. Also, in ecological terms, a population of the three-stage mutant flukes will be able to persist with a lower density of any one of the hosts than can the two-stage flukes. Compare two people, one of whom will eat two food items from a grocery store, and the other who will eat three. Supplies at the grocery store fluctuate up and down, such that sometimes an item may not be stocked. Who is going to starve first? I guess Darwin had arguments of this sort in mind as he wrote the notebook passage quoted above. The evolution of complexity may therefore be open-ended.

Both theories have respectable internal logic. It is almost certain that the complexity of life has a lower limit, which has been occupied for 4000 million years. It is plausible that complexity evolves up and down at random, above that fixed lower limit. It is at least possible that the species at the top of the complexity range will tend to evolve increased complexity, pioneering into the unoccupied space. I doubt whether we can get much further than this by means of argument alone. We need an empirical test. Dan McShea has suggested one way the test could be done. He has been interested in whether complexity tends to increase in evolution, and he proposes that we measure in several lineages the complexity of each species and their descendants in the fossil record. Complexity clearly cannot be measured in these fossils by gene numbers, but we can measure the fossil's structural complexity somehow. We can then simply count whether increases or decreases in complexity are more frequent. He and his followers have done some work, but not particularly to test between the 'pioneering' and 'random' theories. To test them, we should need to break down the fossil species into two sets: those near the top of the range of complexity, and all the rest excluding the ones at the bottom. Then we could measure whether changes up were more probable in the one set and changes up and down equally probable in the rest. It would be hard to squeeze these results out of the published material.

Moreover, the published research has concentrated on relatively short-term trends within fossil groups; W. B. Saunders and colleagues, for instance, have recently described the results of an impressive study of trends in 475 ancestor-descendant lines of ammonites. Ammonites are extinct now; they were a group of shellfish that once thrived in the oceans. The trend in complexity was almost like that in Figure 3c, supporting neither of the main theories I have discussed here. That may place a question mark over the theories, but the study itself may have too small a scale to test them properly. The timescale of the research was, in a sense, huge – 150 million years, or so. But there were two 'ups' and two 'downs' in complexity during that time, and the up trends lasted 50 million years or less. Over that time span, additional factors could operate, factors that would be too minor to notice in a theory about the 4000-million year trend for the whole history of life. For instance, the external environment of the ammonites might have been changing in a way that favoured an increase in the complexity of all ammonites. The environment might show a directional trend for a few tens of millions of years, but it is unlikely that the whole history of life has been driven by a monolithic directional trend in the external environment. In all, McShea has invented a method to test between the theories about complexity. The initial results hint that things may be more complicated than the two big theories admit; but the question remains open for now.

The main purpose of the 'random' and 'pioneering' ecological theories in this book is to show that we can make at least hypothetical sense of the history of live complexity. In Chapter 1 I argued that complex life is a puzzle, even a paradox. The ecological theorizing here provides part of the solution: complex life evolved because ecological opportunities existed for it. But it is only part of the solution, not all of it. I have argued that the more complex life forms were able to evolve by putting more genes to use than did their ancestors, and that the extra genes were in turn supplied by some unrelated process, such as restoring fertility in a hybrid. What I have not yet owned up to is that it is not obvious that extra genes can be added to the existing supply of genes, in such a way as to increase the amount of live complexity that is coded for. Mechanisms exist to add genes, but it is another question how much use these genes can be put to in the future. Other factors can act to limit the amount of genetic information that an organism can contain. The ecological opportunity may exist, the raw material may be furnished by gene duplication, but evolution may still not happen. As we shall see, copying errors in DNA and genetic selfishness can place upper limits on complexity. The evolution of complex life is not only a story about ecological opportunities and multiplying genes. It is also a story of recurrent breakthroughs, as new life forms have evolved and blasted back the limits on complexity. The story will enable us not only to see how complexity arose, but also to make sense of some of the most fundamental properties of life on Earth.

3
The mutational meltdown

Ill fares the land, to hastening ills a prey,
Where change accumulates and genes decay.

Life is built from coded instructions written in the genes, the DNA. When we reproduce, our DNA is copied; and when our DNA is copied, mistakes (that is, mutations) are sometimes made. The question I am going to look at is how these copying mistakes affect the coded instructions, or messages, in our genes. A good analogy is provided by the children's game called 'telephones' in North America and 'Chinese whispers' in Britain. It is the game in which a message is repeated from one person to the next, down a line. By the end of the line the original message is laughably corrupted. The message is corrupted no matter how good the children are at repeating and listening to it, because it is humanly impossible to communicate a message many times without a mistake. The analogy is probably obvious, but to spell it out – the message is like a length of DNA; repeating the message is like reproducing; and mistakes in the repetition are like mutations. The loss of the order, or meaning, from the message by the action of mutation is a simple example of I shall call (following the Oregon school of Michael Lynch and others) a 'mutational meltdown'.

Two features of the game combine to make the meltdown in meaning inevitable. One is that the game is multigenerational: the message is repeated not just from one child to another, but is repeated again and again down a line of senders and receivers. Ordinary human conversation between two individuals works satisfactorily most of the time because minor mistakes of communication do not completely corrupt the sense. Major mistakes do happen and when they do communication fails, but they are relatively rare. In Chinese whispers, errors accumulate down the line. Once one error is made, the newly erroneous message is the starting point from then on; new errors are added to the old

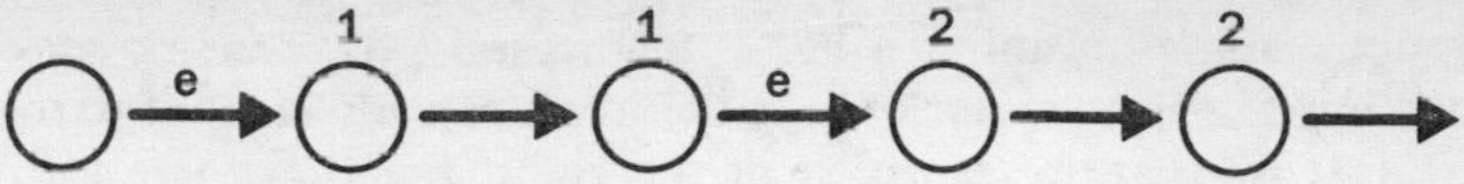

Figure 4. Accumulation of errors. The circles symbolize players of Chinese whispers; the numbers above the circles are the number of errors (e) that have accumulated in the original message.

(Figure 4). If the early errors are minor, the message will still initially make sense, just as a body can survive with a few minor genetic defects. But as the errors accumulate over time they will come to do as much damage as a major mistake. It ultimately makes no difference whether the mistakes are major or minor: the magnitude of the mistakes affects only how long it takes for the conclusion to be reached. The conclusion is that the message is corrupted.

The multigenerational nature of the game also means that it ultimately makes no difference how often mistakes are made. If mistakes are rare, the message will decay at only a trickling rate; if they are common, it will decay more rapidly. But at any rate, ultimate corruption is inevitable. Whether the ratchet of accumulating error turns once every two repeats, or once every twenty repeats, it will turn and turn until the message is mangled out of existence.

The errors are essentially irreversible. Once a mistake is made, no amount of subsequent accurate repetition can make up for it, because the message that is now being accurately repeated is wrong. A second mistake downstream from the first mistake could theoretically reverse the first mistake and restore the original sense. But this reversal would be but a temporary blip in the inevitable decay. If you take a corrupted message and introduce another error into it, the balance of probabilities is such that you are unlikely to restore the original. The balance is against you because there is only one way of being right (the original message) and so many ways of being wrong (anything other than the original message). Suppose the original message was a speech from *Hamlet*: some mistakes have been made and we now have a partly corrupted speech; parts of it are changed from the original and parts are unchanged. Further mistakes are now introduced. If they are in the unchanged parts, they necessarily deepen the corruption; if they are in an already changed part they probably substitute one corrupt reading for another. For a new mistake to reverse a previous mistake it must not only strike in an already corrupt part, but also, by chance, change it in the right direction to restore the original. It is not exactly true that the number of errors in the message inevitably increases as it is repeated, but over time the number does tend ever upwards and the ultimate corruption of the original message is overwhelmingly certain. This is an uncontroversial claim. It is no secret what happens in a real game of Chinese whispers.

The second disastrous feature of the Chinese whispers game is that a child repeats the message only once. Living creatures produce more than one offspring, making several attempts at repeating their DNA messages. Natural selection can then operate. Some of the offspring produced by a parent may be free of mutational errors – the other offspring, who contain mutations, will be inferior because of these mutations and may die without breeding. The DNA can be passed on to the next generation without any accumulation of error, because only the error-free offspring breed. The original message can now be perpetuated indefinitely, despite the copying mistakes. Natural selection can immortalize a message.

In the game of Chinese whispers, if each child repeated the message to several (say, five) other children, the original message could be maintained, uncorrupted, down an indefinitely long line, provided that there is an error-free version at every stage (Figure 5). The error-free version does have to be selected. How is this selection achieved? A supervisor might listen to all five repeats and pick a player who received an error-free version to continue and repeat the message in turn. But readers with a more naturalistic, or more diabolical, imagination will see that the supervisor is unnecessary. For instance, the initial message might contain some good advice about how to stay alive, such as WHEN THE TIGER COMES, FREEZE (I do not know whether it helps to freeze with tigers; let us assume it does). Then, from time to time, tigers can be released into the room, and the children who have received an uncorrupted version of the message will be more likely to survive. Alas for the children who heard WHEN THE TIGER COMES, SNEEZE, or TEASE, or SQUEEZE. We are now close to the way in which natural selection preserves order in real DNA over the generations. The message is an instruction, like a genic instruction in a live body. The particular instruction about tigers

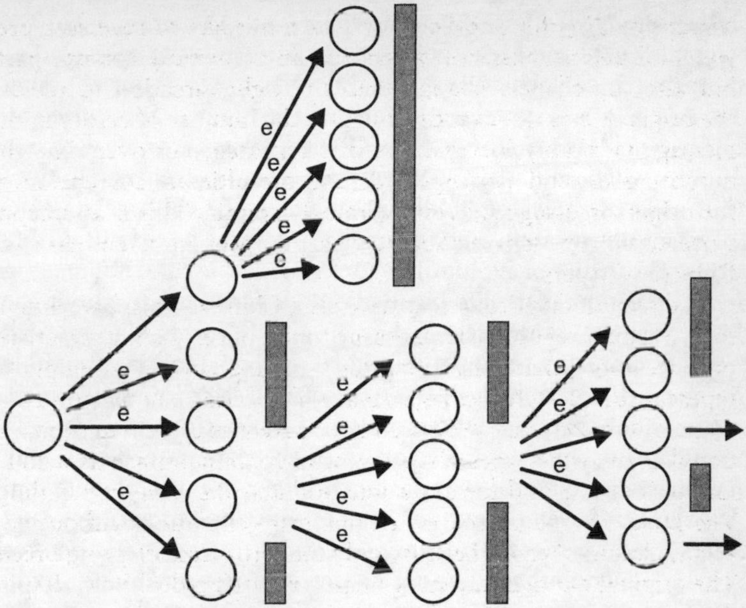

Figure 5. Extended Chinese whispers game with multiple repeats. Each player repeats the message to five receivers. Only receivers who have an error-free version of the message are allowed to repeat it. The message can now be perpetuated indefinitely without error. The message is preserved in this game in a manner analogous to natural selection in life.

would be represented in the DNA code in a less obvious way than a gene coding for a digestive enzyme or a skin pigment. But the genes can influence our brains. A gene might, for instance, cause a connexion to develop between the nerves that recognize tigers and the nerves that cause behavioural freezing. This subtle alteration of brain wiring would correspond to the instruction verbally represented in WHEN THE TIGER COMES, FREEZE.

The message of a Chinese whispers game is short, but it could be elaborated into a multi-instructional message to meet the innumerable demands of life. When the tiger comes, freeze; when the dinner comes, tuck in; when the attractive member of the opposite sex comes, strut your stuff. I have moments of imagining life as an elaborate game of Chinese whispers, with multiple receivers, in which the message that ends up being preserved is one describing how to survive the horrors of the environment where the game is being played. A message about how to survive is the only kind of message that can achieve immortality. A message about anything else, such as about survival on Mars, will be rapidly randomized by the same process as in the original Chinese whispers game with one repeat: it will be mutated away, with no force to prevent corruption. Martian know-how does not help you against Earthly tigers.

The argument allows us to predict a general feature of all life forms. The theory of evolution is sometimes accused of not making predictions; but that says more about the critics than about the theory. Here we can predict that, in all life forms, parents will produce more than one offspring. If each parent in its whole lifetime produces only one offspring, natural selection is powerless to prevent the mutational meltdown. (I mean a *potential* of only one offspring. In all species, many of the parents produce only one offspring, but one from a much larger number of potential offspring; and even if many parents produce one offspring, many others produce more than one, or zero. It is automatic one-to-one reproduction that rules out natural selection.) Mutations will over time inevitably randomize any living organization. Actually there are no known life forms in which individuals routinely produce one offspring. We can also predict that none will ever be found here or anywhere else in the Universe. A few mythical creatures mythically reproduce in a one-to-one manner. The phoenix, for instance, is reborn in the ashes in what is essentially a one-to-one reproductive event. And as predicted, phoenixes do not exist.

In the children's game, the meltdown of meaning is merely humorous. The original message will change into some other message: maybe it is nonsense, maybe not, but it hardly matters. Nothing big has happened, the world goes on, the players go and do something else. But in life – this is the point of the analogy – the same process is deadly serious. Life is not a party: DNA cannot accidentally change many times from one sequence to another and life go on as normal. The original DNA would have been coding for a living creature; but if errors then accumulate in it, it will soon be coding for a non-living creature. If life were logically like Chinese whispers, life would not exist.

The message in the game is corrupted because the one original message is a tiny minority of all possible messages. The DNA sequences that support life are also a tiny minority of all possible DNA sequences. We can be sure of this because live systems possess what biologists call adaptations. Living creatures are designed (that is, adapted, in the sense of fitted or suited) for staying alive. We have eyes for seeing, stomachs for digesting, livers for detoxifying, hearts for pumping blood. Eyes, livers and hearts are astronomically improbable structures; they do not just spontaneously form if you jumble up chemicals in a test-tube. Eyes, livers and hearts are ultimately built using instructions in the DNA, and those instructions are not any old random sequence of DNA; they are the specific sequences that code for instructions that work. It would be a hard task to write out a DNA sequence, made up of the four letters A, C, G and T, that coded for a live creature. If I wrote out some sequence of those four letters, it would not code for life. They would definitely not code for life if I wrote the letters randomly, but I could not write a sequence of DNA that coded for a life form even if I tried to – I do not know enough about the molecules of life. In the world today there is probably nobody who could write out a DNA sequence coding for life, and even if someone could, it would be for only a feeble specimen of life. A degenerate parasitic bacterium that can only survive while plugged into someone else's life support system uses a thousand genes, or about a million nucleotide letters. There are $4^{1,000,000}$ possible sequences that long. I do not know what fraction of them code for a live bacterium, but I do know that the fraction is strikingly small.

The same probabilistic reasoning underlies another principle of this and the next two chapters. We shall be concentrating on mutations that make things worse and be ignoring the perhaps more familiar class of mutations that allow evolutionary change. We can divide mistakes into three classes: advantageous mistakes, which make things better; neutral mistakes, which make no difference; and harmful mistakes, which make things worse. The first two kinds of mistake underlie all evolutionary change in the world. When HIV evolves to become resistant to a drug, it is because of an (advantageous) mutation that allows the virus to evade the drug. Advantageous mutations underlie adaptive evolution. This virus also changes over time because of neutral mutations. Neutral copying errors appear in the viral genes, and the mutant forms drift up and down in frequency over time in the viral population. However, our topic here is not evolutionary change. Our topic is mutational decay, and how life preserves itself against it.

The decay of a message depends mainly on the rate of harmful mutations. We can ignore the advantageous mutations, for two reasons. One is that they are rarer than harmful mutations. A random mistake in the DNA is about as likely to improve the creature it codes for as random change in *Hamlet* is to improve the play. The frequency of advantageous mutations is so small that we could ignore them. Alternatively, we could be more exact and define the rate of harmful mutations as a net figure, after the few advantageous mutations have been accounted for. The net rate of harmful mutation equals the total rate of harmful mutation minus the advantageous mutations. We can either be informal and ignore the advantageous mutations, or be formal and define them away. We can also ignore the neutral mutations, for another reason: not because they are rare but because they are irrelevant. A neutral mutation is like a 'mistake' that does not alter the message. A neutral mistake could be something like a spelling variant, as if the speaker imagines the word as TYGER and the receiver imagines it as TIGER. It is (in a sense) not a mistake at all, and in tracing the decay of the message we can simply look through the neutral mistakes as if they were not there. We are left with the harmful mistakes as the only mistakes that matter. We should need to think about the other kinds of mistakes if we were thinking about some other topic, such as the molecular clocks that are used in evolutionary timekeeping, or the creative evolution of adaptations such as the eye. For the mutational meltdown they are two things we can ignore.

I have also simplified the problem by treating the anti-tiger message as if it were permanently useful. In practice, the message that a life form needs will change over time as the environment changes. Tigers, for instance, may go extinct or evolve new hunting techniques. In a million years' time it may be better to put your fists up, like a boxer, when you see a tiger. If the message that is needed changes over time, it slightly alters the chance that a mistake is advantageous as opposed to disadvantageous; but the same basic problem exists. It will still generally be true that a mistake is much more likely to make things worse than better.

Analogies arouse suspicion, but Chinese whispers is actually a powerful way to think about mutation. Here are two differences between the game and life that might be thought to matter but turn out not to. The first is that the game has only one reproducing lineage, whereas life has a whole population of parents copying their DNA into offspring every generation. But it turns out that the same meltdown happens in a population as in any single line. Imagine a house with several rooms, each containing a line of Chinese whispering children. If all the lines start with the same message, it may decay at different rates in the different rooms, but it is going to decay in all of them eventually. The sense will be lost from one line of messages after another, until it is lost from them all. It therefore makes little difference whether we think about one line, or a population of lines. The other difference between the game and life is that a human receiver may smooth out a mistake in the received message, such that the message is still made up of real words and still makes some kind of sense. The message will then not be corrupted into a random series of syllables, and it may retain a meaning that wanders around over time. A message might even be improved in some respects as it was repeated between human beings. I am thinking of poems – Homer's possibly – that may have been partly invented in oral tradition. But the conclusion that matters is not altered: the original message will be lost. Meaning is free to wander around only when it is not vital. If the message were really about how to survive tigers, then change would become as deadly as mutation in DNA. The best analogy for DNA is therefore a tiger survival message, not a poem or theatrical speech. Moreover, errors in DNA replication are smoothed out in a manner not unlike a human receiver. There are only four kinds of nucleotide letter in DNA, and mutations are more or less random among these four (although some changes are more frequent than others); but mutation is still usually constrained to generate some sort of DNA message, rather than producing a random combination of carbon, nitrogen, oxygen and hydrogen atoms. We can ignore the detailed differences between Chinese whispers and the copying of DNA. The analogy points to a property not just of children's games, but of life too.

The ceiling on complexity

If every parent produces only one offspring (or, in Chinese whispers every individual repeats the message to only one receiver), the message will be corrupted and natural selection cannot stop it. Any errors are too many. With multiple offspring, natural selection can help – but even in this case selection can maintain order only if the error rate is not too high. How high can the error rate go now? Figure 5 showed the multiple-repeat Chinese whispers game with five repeats per child. If the error rate is 1/5th then about one of the five repeats will contain an error and four will be error-free. Selection can easily maintain order. Any of the four error-free children can be used to perpetuate the message, which can be transmitted perfectly from one generation to the next, generation after generation, for ever. If the error rate were somewhat higher, at 3/5ths, then more like three children will have bungled messages, but the message can be perpetuated by selecting one of the two children with error-free messages.

We reach the theoretical upper limit when the error rate is so high that only one of the five repeats is correct. The message can theoretically be passed on indefinitely provided that there is one error-free repeat per generation, because that one child can form the link to the next generation. But if the error rate exceeds this upper limit, and sometimes (even if very rarely) every repeat in a generation contains mistakes, selection will be overwhelmed. Selection cannot maintain order if all the offspring contain more mutations than their parents (Figure 6). Each such generation irreversibly corrupts the message. If the error rate is high enough for a generation like this to happen once, it is high enough for another to happen later, and it is only a matter of time before the message will be randomized. Decay over time is inevitable. When the error rate is high enough for every repeat to be mistaken, it no longer really helps to have multiple receivers. Selection can at best only slow the progress of corruption (by selecting the least bad repeats) but cannot halt it. A mutational meltdown proceeds much as in the original game with one receiver.

With one offspring per parent, natural selection cannot maintain order if any errors are made at all; the mutational meltdown proceeds under the pure influence of mutation. In real life, with multiple offspring per parent, selection acts as well as mutation:

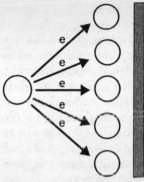

Figure 6. Chinese whispers, with the same extension as Figure 5: a player repeats the message to five recipients and only those with an error-free message are allowed to repeat it. Now the error rate is so high that all recipients have an erroneous message and natural selection puts a stop to them all (alternatively, a recipient with the least erroneous message might be allowed to repeat it, and the message would be randomized in the same way as in Figure 4). Life cannot exist with an error rate so high that all offspring contain new mutations.

and at any moment the forces of mutation (which introduces error) and selection (which removes it) act in opposite directions. If selection removes more errors than mutation introduces, the quality of life improves. If the number of errors introduced equals the number destroyed, the balance is exact and the quality of life is constant. This balance is presumably the condition of real life, at least if we average over a sufficient time period. If the number of errors that mutation introduces exceeds the number that selection removes, then the system is in a mutational meltdown and adaptation is being lost. Natural selection could keep the population pure only by killing all the offspring. The population either goes extinct instantly, if natural selection kills everyone, or slowly, if some of the offspring survive and the order in the DNA that supports life is randomized away. Either way, the life form will not exist for long. The upper limit on the error rate in the multiple-receiver game is that at least one of the recipients has to have a perfect message in every round of repetition. If this error rate is exceeded, the tigers have the feast that ends all feasts.

In the game (with multiple receivers), the need for at least one error-free repeat is absolute, not an average. In life, however, the theoretical upper limit can be relaxed to an average: life needs an average of at least one error-free offspring per parent. The reason is something we met before, that life has many reproducing lineages, as if Chinese whispers were played simultaneously by many lines of children in separate rooms of a house. Last time we met this difference, it did not matter; this time it does. In a population of message-repeaters, accurate repeaters can make up for blundering repeaters. Suppose each child repeats the message five times and the average for the house as a whole is for a child to manage one accurate repeat and four erroneous repeats. In one room a blockhead makes mistakes in every repeat; there the tigers scoff the lot. But if the average is one perfect repeat per repeater, someone in another room has probably repeated the message without error to two recipients. The tigers go hungry there, and the quality of the message in the house as a whole is therefore maintained. (You can even imagine that where fewer children are eaten, things become a little crowded and some of them wander through the house looking for space.) In life, the upper limit on tolerable error is also set by the condition that every parent must on average produce one error-free offspring. If the life form is one that takes two parents to make an offspring, then every pair of parents must produce an average of two error-free offspring. Any more error than that and we are not talking about life.

We can now pick up again the big theme of the book, the evolution of complexity. Complex life forms contain more genes than simple life forms; their DNA messages are longer. A complex life form is more likely to make a mistake in copying its DNA for the same reason that you are more likely to make a mistake when you repeat a Shakespearean soliloquy than when you repeat a snappy advertising slogan. A virus is rather like a slogan; it is short and easy to copy. We are more like the complete works of Shakespeare. The evolutionary preservation of human beings poses a problem rather like the preservation of Shakespeare's complete works if it were transmitted orally, without written copy. There would need to be something very special about the people who did the transmission, or Shakespeare would soon be lost. As DNA messages have grown longer, in the evolution of increasingly complex life, copying mistakes have become more of a problem. Live complexity hits its ceiling when the DNA message is so long that a mistake happens every time it is copied. When life is near this upper limit, complexity cannot evolve upwards even if there is an ecological opportunity there.

It might be argued that one mistake in a long message is likely to be less damaging than one mistake in a short message, and complex life could tolerate more errors than simple life. The extra mistakes in a longer message would be balanced by the lower harm per mistake. One or two typos in a book is no disaster; one or two typos in an advertising slogan is. A syphilis bacterium with one gene mutated in its set of a thousand genes may not be able to cause syphilis, but a fish with one mutated gene in its set of a hundred thousand genes may be still able to spawn and swim. That may (or may not) be true, but it does not matter because the amount of damage that a mistake causes is relatively unimportant. What matters for the mutational meltdown is the number of mutations: the meltdown happens if the number of mutations that arise is bigger than the number that natural selection removes. If the error rate is so high that every offspring contains an error, it ultimately does not matter whether the error reduces the offspring's chance of survival by 1/10th or by 1/10,000th. The errors are accumulating, and will continue to accumulate until the DNA is randomized and the species is extinct. The damage done by an error influences the rate at which the quality of the life form declines, but not its existential possibility. We need to distinguish between the survival of an organism for its lifetime and the existence of a whole life form, or species, in evolution. An organism is more likely to survive if 1/100,000th rather than 1/1000th of its genetic instructions are knocked out. But a life form cannot exist if every generation an extra 1/100,000th (or any other positive fraction) of its genes are randomized. The crucial question is whether the error rate is so high that all the offspring contain errors; the damage per error is a secondary matter and can be ignored.

The uses of redundancy

Life has many methods of dealing with error. Errors can be prevented from happening to begin with; they can be corrected; they can be disguised; they can be purged by natural selection (Table 2).

Table 2. Factors that influence the number of errors (or mutations) and the efficiency with which natural selection removes them. The final entries of the two columns are two ways of saying the same thing.

Error	*Selection*
Total length of DNA message	Number of offspring per parent
Unit error rate	Fraction of death claimed by other mortality factors:
Fraction of errors that do (or do not) do damage: (i) number of repeats of message (ii) other ideas	(i) accidents (ii) other forces of selection
Even, random, or scapegoat distribution of errors among offspring	Number of errors removed per death

Life can evolve to become more complex if it improves any of these methods. One way to hold the total error rate down is to reduce the unit error rate – the error rate per unit length of the message. For a verbal message, it is not obvious what a 'unit' is; maybe it is a syllable. For a DNA molecule, the answer is obvious – it is one nucleotide letter, such as A, C, G or T. In Chinese whispers, the players could reduce their error rate by listening more carefully, repeating the message more slowly and methodically, or by having a third person check the message as it was repeated. Life has analogues for most of these, and we shall look at some details in the next chapter. Here I want to concentrate on a more abstract feature of DNA.

The DNA molecule is a double helix. The codes of life are the sequence of letters, such as ... CATTACA ... but the codes are carried in double form: the coded instructions themselves, together with another complementary sequence of letters in the other strand of the double helix. Each letter is chemically bonded to another complementary letter in the other strand: A bonds to T, C bonds to G. The sequence CATTACA in one strand will be matched by the sequence GTAATGT in the other strand. It is as if a child passes on the message WHEN THE TIGER COMES, FREEZE not once but twice, once in the straightforward form and a second time in some complementary version, such as EZEERF SEMOC REGIT EHT NEHW. The important abstract feature of the doubled message is that it contains redundancy. Mistakes in one version can be corrected using the other version.

The mistakes can be detected as mismatches. The sequence CATTACA should have GTAATGT lined up next to it. But a T may have been miscopied, giving GTAGTGT. We need to distinguish mistakes that happen during copying from mistakes that happen spontaneously within the message at other times. I have been treating errors as if they were all transmission errors, when one child passes the message on to another. But an error could also arise within one child, as if each child had to remember the message for some time before passing it on. The DNA is vulnerable to spontaneous decay between copying events. The redundancy inherent in the double-strandedness of DNA is no advantage in correcting copying mistakes, but it is useful in correcting mistakes that happen in the double helix at other times. Suppose a C in one strand of the DNA mutated to a T. There is now a mismatched T–G pair in the two strands, where there was a good C–G pair before.

If you are just shown a T–G pair you do not know whether the T is right and the G wrong or vice versa. The double-strandedness of the molecule does tell you that there is something wrong, but it does not tell you what to do about it. Even this can be an advantage: if you had a single-stranded message in which a C had mutated to a T you would not even know there was a mistake. But the big advantage of the redundancy comes when a letter spontaneously changes to something other than one of the four authentic letters of DNA code. The C might mutate to something else completely, which we can symbolize by X. We now have an X–G pair instead of a C–G pair. In a double-stranded molecule we know that G is right and X is wrong, and the X should be replaced by a C. In a single-stranded molecule we should just have an X and not know what to correct it to. The redundancy inherent in the doubled-up codes of DNA is a powerful device for detecting copying error, and

69

reconstructing the correct code from the complementary strand.

Redundancy is a standard method of dealing with error in human communication. We trade off the economic advantage of a succinct message against the danger of error, and human communication would be impossible without its obvious repetitions and less obvious redundancy. Compare, for instance, 'eighteen filthy nine' and '18T9'. In the longer, verbal version we can easily reconstruct the correct signal because of the inherent redundancy in using a five-letter word ('fifty') to stand for the digit '5'. The version '18T9' could stand for any of ten dates, and the correct signal is irrecoverable. The short version of the signal may be cheaper to transmit, but it is more vulnerable to error.

The double helix is not the only form of redundancy in the system of Mendelian inheritance that we and other complex life forms use. We carry two sets of double-helical molecules, one inherited from our father and the other from our mother. Our cells contain four copies of the information for each genetic instruction: a maternal and a paternal double helix. This is a condition called 'diploidy'. Ploidy refers to the number of DNA sets that an individual inherits. We, like almost all familiar forms of life, inherit two sets of DNA. This contrasts with many kinds of microbial life, such as many bacteria, which are 'haploid'. Bacteria inherit only one DNA molecule. If we ignore the complementary set due to the double helix, then a bacterial message states each piece of advice only once: WHEN THE TIGER COMES, FREEZE, and that is all. In human beings, the genetic instructions say: WHEN THE TIGER COMES, FREEZE peep WHEN THE TIGER COMES, FREEZE. Some species are triploid, or tetraploid, or polyploid, in some cases multiplying up the number of DNA sets to apparently crazy levels (WHEN THE TIGER COMES, FREEZE peep WHEN THE TIGER COMES, FREEZE peep [repeat n times]).

Why do we have two DNA sets? It is not logically inevitable. Sex does require a doubling and halving of the gene numbers through the cycle of life and reproduction. Otherwise the number of DNA sets would increase to infinity or shrink to nothing. The doubling does have to happen at conception (or fertilization, in more general biological terms). But the halving could logically happen at any time after conception. We keep our two sets for almost all our cells throughout our lives. We only halve the gene numbers, at the last minute, to produce the sperm and egg cells. In other life forms, individuals halve their gene sets almost immediately after fertilization and have one set of genes per cell for almost all their lives. Moss and the single-celled creature called *Chlamydomonas* are examples. Why do we allocate almost 100 per cent, rather than some other fraction, of our lives to the two-gene set rather than the one-gene set phase?

One tempting answer, given what we just saw for the double helix, would be that we keep the second set to correct errors in the first. Maternal errors could be corrected using paternal codes, and paternal errors using maternal codes. The offspring would be fine, as long as it did not inherit errors in the same bit of DNA from both its mother and its father. However, the two gene sets in diploid creatures are not used in this way. The reason is probably that it is impossible to tell which is right and which is wrong. If your mother says WHEN THE TIGER COMES, FREEZE and your father says WHEN THE TIGER COMES, FLEE, how do you know which to believe? It was possible to correct errors in the double helix because the two strands lie side by side and we knew which to trust when one strand had an impossible letter. But those mistakes have all been corrected by the time the paternal and maternal messages are put into the offspring. Both messages will be plausible in themselves.

However, the redundancy of our diploid gene sets may still help. It may, in the right circumstances, allow us to override a bad version of an instruction with a good version. We can stay with the standard example.

Original message:
    WHEN THE TIGER COMES, FREEZE

Now imagine three kinds of mutant message.

Mutant 1:
    WHEN THE TIGER COMES, FLEE

Mutant 2:
    [silence of five words' duration]

Mutant 3:
    WHEN THE TIGER COMES, FLEE AND DO NOT FREEZE;

MESSAGES ABOUT FREEZING ARE ENEMY PROPAGANDA
AND SHOULD BE IGNORED

A bacterium, with one copy of the message, might inherit any of these three kinds of error, and it would be worse off with all of them. But we inherit two sets, and usually at most only one will contain an error in any one instruction (because errors are rare, and the maternal and paternal genes are copied independently). Now the original and the three mutants are as follows.

Original message:

WHEN THE TIGER COMES, FREEZE. WHEN THE TIGER
COMES, FREEZE

Mutant 1:

WHEN THE TIGER COMES, FREEZE. WHEN THE TIGER
COMES, FLEE

Mutant 2:

WHEN THE TIGER COMES, FREEZE. [silence of five words'
duration]

Mutant 3:

WHEN THE TIGER COMES, FREEZE. WHEN THE TIGER
COMES, FLEE AND DO NOT FREEZE; MESSAGES ABOUT
FREEZING ARE ENEMY PROPAGANDA AND SHOULD BE
IGNORED

The important case is mutant number 2. Here the good version will override the bad version. Doubling up the gene set is a big advantage for these mutants. Doubling up the genes is pointless for mutants of type 1 and actively stupid for mutants of type 3. But biological mutations are more often of type 2 than of type 3 (the unrealistic nature of the third mutation is suggestive here). Doubling up the gene messages is probably, on average, advantageous in life. We therefore contain two kinds of redundancy. One, in the double helix, allows errors to be corrected; the other, in the form of diploidy, allows errors to be masked.

In Table 2, diploidy works by reducing the fraction of errors that do damage. It reduces the error rate by disguising, rather than avoiding or correcting, errors: it acts to neutralize harmful mutations. Other factors may also influence whether mistakes are neutral or harmful. One is the amount of rubbish that the message contains. If a message is 10 per cent meaningful and 90 per cent noise and you introduce a random error into it, the 'error' will probably be in the meaningless part and will not matter. If a message is 100 per cent meaningful, the same amount of error is much more likely to make things worse. Anyone who has marked examination essays knows how difficult it is to spot errors if they are from within a region of text that is already mainly rubbish. Whether examination candidates write rubbish for this reason is another matter. Live species probably have not evolved rubbish (or junk, as it is more often called) in their DNA to dilute their mutational errors. But some may have evolutionarily acquired junk for other reasons, and we need to take the junk into account in thinking about the error rate that a species can tolerate. Living creatures do not need to be careful about copying rubbish.

Zombies and scapegoats

Whether or not a life form can exist depends not only on its average copying accuracy, but also on how erratic the copying is. Suppose, for example, that the average error rate is one per off-spring. Perhaps every offspring has exactly one error, in which case the life form would not exist. Or perhaps in a brood of ten, nine have no errors and one has ten errors; all the errors are concentrated as if in a scapegoat. This life form could comfortably exist – natural selection has nine error-free offspring per parent to play with – despite having the same average error rate as the previous biologically impossible case. Or the errors could be random, which gives a distribution somewhere between the even and the scapegoat extremes. A brood of ten offspring with randomly distributed errors, occurring at an average rate of one per offspring, contains about two or three error-free offspring. Biologists tend to assume that mutations are randomly distributed, but the assumption is only reasonable rather than certain.

The interesting possibility is that errors are concentrated in a minority of scapegoat offspring. A life form can exist despite a high error rate if it concentrates its errors rather than scattering them at random. Imagine someone who is repeating a message to a number of receivers (Figure 5). This person could be rather zombified, concentrating exactly equally at every repeat, taking equal care to say the message clearly and correctly at every repeat. Occasional errors would slip through, at random, but every recipient would be equally likely to receive an error. Alternatively, the message-repeater might prefer to concentrate really hard on the repeats for a minority of recipients, at the expense of giving less attention to the others; then some recipients would be more likely than others to receive erroneous messages. The extreme of this concentrate-and-relax procedure is the scapegoat distribution, in which a few unlucky recipients get all the blunders. Whether message-repeaters used one method or the other in the game would depend on how they could allocate their mental effort most efficiently: some people have their ups and downs, others are steadier performers. There could be analogies in DNA copying, for instance if an individual could copy its DNA better under some conditions – such as when it was well fed and relaxed – than others.

The same process could operate at the level of the population. We thought earlier about a house containing several rooms, with a row of children playing Chinese whispers in each. They are repeating the same initial message in every room. It could be that in some of the rooms the children are all sick, or do not want to play, or it is the noisy side of the house and hard to hear the message. Then in the house as a whole, some recipients are more likely than others to be told an erroneous message. In life, some individuals may be particularly bad at copying their DNA, perhaps because they have defects in the DNA-copying machinery. But whatever the exact reason is, we need to take account not only of the average error rate but also how the errors are distributed among all the offspring in the population. What ultimately matters is not simply the average error rate but the ability of life to produce enough error-free offspring.

The death sentence

We can now turn to the right-hand column of Table 2 to consider the ability of natural selection to purge error. Fecundity (the average number of offspring per parent) is one possibility. We could have the players in Chinese whispers repeat the message five times in one place, but fifty times in another. The more times the message is repeated, the more likely it is to be repeated correctly at least once. The fifty-repeater children could perpetuate the message, and hold out against the tigers, down a longer chain than the five-repeater children. For the same reason, a life form in which each parent produces five offspring could only persist with a lower error rate than a life form in which each parent produced fifty offspring. In the history of live complexity, the number of offspring per parent seems to have been traded off against other factors; human beings and other brainy animals produce small numbers of offspring per parent. But other things being equal, more errors can be tolerated if more offspring are produced.

A second factor is the number of errors eliminated per death. Natural selection works by causing the premature death of individuals, and the more errors there are in those individuals who die, the more error natural selection removes. Imagine a population in which fifty individuals out of a hundred breed and perpetuate the population, and the other fifty die without breeding. If the individuals who die contain one error each, natural selection has purged fifty errors, or 0.5 of an error per average member of the population. If the individuals who die contain ten errors each, it has purged 500 errors, 5 per average member of the population.

Ploidy – the number of sets of genes in an individual – can influence the numbers of errors removed per death. One of the mutations we thought about said 'WHEN THE TIGER COMES, FREEZE. [silence of five words' duration]'. The error is masked by the unspoiled version of the message, and the mutant individual can survive as normal. The error is purged only when it appears in double form: '[silence of five words' duration] [silence of five words' duration]'. The tiger who eats this individual will be digesting two mutant genes in one dinner portion. Compare that with a life form that has only one gene set. When the same mutation occurs, the individual's message reads '[silence of five words' duration]'. It is as likely to be eaten as the double mutant, but its death will remove only one copying error. When life evolved from haploid to diploid, doubling its gene set, it not only came to disguise some of its copying errors, but also to enhance the power of selection. Selection, for a certain class of mutations, now removed twice as many mutations in every death.

The distribution of errors among the offspring can also influence the power of selection. If all the error is concentrated in a minority of scapegoats, many errors can be cleared in one death. But if the errors are evenly distributed, and all the offspring have a similar number of errors, it takes much more death to clear the same amount of error. In general, a life form can survive with a higher error rate if it has some mechanism that allows natural selection to eliminate many errors at a time.

The efficiency of natural selection also depends on what other causes of death are at work, in addition to copying error. A life form that produces an average of one error-free offspring per parent can perpetuate itself indefinitely in theory. But in practice it can only do so if genetic error is the only cause of premature death. A species in which each parent produces ten offspring, of which nine on average contain mutations and one is error-free, is at the limit of the possible. If the nine mutants die and the one error-free offspring survives to reproduce, the species will persist with no genetic deterioration. But it is almost impossible that death will visit with enough genetic precision to keep this species out of a mutational meltdown. Too many other things can influence whether or not an individual lives or dies. In particular, I want to divide causes of death into three categories. A creature may die because of (i) a copying error in its genes, (ii) an accident, or (iii) some other kind of natural selection – that is, other than (i). The theoretical upper limit on the error rate assumes that the second and third factors are inoperative. This assumption is not realistic.

By accidents, I mean causes of death that have nothing to do with genetic differences between individuals. On All Souls Day in 1755, an earthquake struck Lisbon and the cathedral collapsed, killing the people inside. Was the death selective? Voltaire provided a famous Voltairean analysis, in which the Christians, being inside the cathedral, were more likely to die, and the practising atheists outside were more likely to live. If some people had a genetic tendency to be Christian, or assemble in large buildings on a holy day, then selection would be at work. If there were no such genetic tendencies, the death would have been genetically accidental. Any particular example can be argued with, but I suspect that earthquakes are mainly genetically accidental. Accidental death surely does happen, and strikes out genes indiscriminately, independent of their copying accuracy. A species in which a parent produces only one error-free offspring could not really survive. The error-free offspring might accidentally be hit by falling masonry, and the species would either not be perpetuated or perpetuated only by error-containing offspring. If a real life form is to avoid mutational meltdown, it must produce not just one error-free offspring per parent but one plus enough of them to cover for accidental deaths.

A similar point applies when natural selection is working not only against error but also on something else. In a bacterial population that is evolving resistance to an antibiotic, some of the bacteria are genetically resistant to the antibiotic and others are genetically susceptible. Natural selection is acting, as the susceptible bacteria are more likely to die when we treat them with antibiotics. The force of natural selection to remove copying errors has to be added to the selection against the antibiotically susceptible bacteria. Again, this creates a problem if the species is operating at the upper limit of the error rate. The species has no room for any other kind of selection. If half the bacteria in the population are resistant, and half are susceptible, there is a half chance that any one error-free individual will be susceptible. The average parent then needs to produce two error-free offspring for there to be one resistant, error-free offspring to carry the system forwards.

The theoretical upper limit on the error rate for transmitting the tiger-survival message was reached when only one child received it error-free at each round of repeats (Figures 5 and 6). This assumes that the one child with the error-free message does not have an accident, and that no other selective factors (e.g. grizzly bears, scorpions, germ guns, death rays) are at work, each with its own survival message. The child who has the right tiger-survival message may not be available to pass it on if he or she had the wrong message for scorpions or went to church on an unlucky day. The upper limit of one error-free offspring per parent is a very theoretical upper limit. It leaves no margin of safety for accidents and other kinds of selection that may remove the error-free offspring. A realistic upper limit on error, for order to be maintained over a long period, is that every parent must breed one error-free offspring plus a comfortable safety margin of additional error-free offspring. This is deliberately vague, because no one knows how much death is demanded by accidents and other sources of selection. But some surely is, and the mutation rate must, therefore, not be so high that the average family lacks a decent number of error-free offspring.

A rough and ready rule about error rates

An average parent must produce at least one offspring without copying error: this criterion sets the formal upper limit on the error rate. For this criterion, we look at the parent's whole brood of offspring and ask whether enough of them are free of errors. But it turns out that it is often easier to think in terms of the number of errors made when producing any one offspring. What is the upper limit on the error rate when expressed like this? The rough and ready answer is: 1. A life form cannot exist if it makes more than one mistake per offspring. We can see that the rule is only approximate. With an average error rate of 1 per offspring, some of the offspring will be error-free and the species is theoretically sustainable. A random distribution of errors in a family of ten offspring might produce something like three offspring with no errors, four offspring with one error, and three with two errors. There are three error-free offspring for natural selection to play with, more than the theoretically required ration of one.

So a species can theoretically persist with an error rate of 1 per offspring. This might suggest that the upper limit on the error rate is more than 1. But we also need a safety margin. With an average error rate of 1 per offspring, there is some safety margin. Its size depends on the number of offspring per parent. With ten offspring per parent, the safety margin is about three: two-thirds of the error-free offspring could die and still leave one to carry the system forwards. With a hundred offspring per parent, the safety margin is more like thirty, and 29/30ths of the offspring could die for accidental reasons without putting the species into a meltdown. In fact we do not know what safety margin is required, and the upper limit on the error rate of 1 per offspring is an approximate figure that takes account of the need to supply some safety margin. At least we can say that the error rate of living systems cannot be much higher than 1. With an error rate of 1, about 30 per cent of the offspring are error-free. If a parent makes an average of 3 mistakes per offspring, the fraction that are error-free goes down to about 4 per cent; and with an error rate of 5 per offspring, only 0.5 per cent of the offspring are error-free. A life form with an error rate of 5 could only persist if the parents produced huge numbers of offspring. We can conclude that parents, in all species, will not make much more than about one mistake on average when they produce a child.

Life can only be as complicated as its rate of copying error permits. Life has to stay on the right side of the mutational meltdown. If the rate at which mutation introduces error exceeds the rate at which natural selection removes it, the life form will mutate itself into oblivion. Mutation rates have an automatic tendency to increase as life evolves to become more complex, because the DNA messages of more complex life forms are longer. We shall see in the next two chapters how the evolution of complexity has been accompanied by new mechanisms to drive down the error rate, or improve the power with which natural selection purges error. At the origin of life, when simple molecules replicated on the face of the Earth, an ecological opportunity probably existed for a more complex form, perhaps cells, that could consume those molecules, or sequester the resources needed for the naked molecules to copy themselves. The opportunity was seized only when catalytic copying enzymes evolved, improving the accuracy of copying and lifting the upper limit on the complexity of the live molecules. Complex life would have remained a multimillion year science fiction if those copying catalysts had not evolved. Further mechanisms arose, and allowed 1000–5000-gene bacteria, 5000–15,000-gene eukaryotic cells, and 15,000–60,000-gene multicellular creeping things to appear on Earth. Maybe, even now, the Earth could support a million-gene life form, twenty times more god-like than ourselves. Life has been stuck at an upper limit of 60,000 genes for 500 million years, and the next breakthrough in Earthly complexity may be waiting for a new invention in the table of error-reducing and error-purging mechanisms.

4
The history of error

Learning from nature's mistakes

When parents make offspring, they also make mistakes. The question in this chapter is how many. We have deduced that a life form cannot exist in which the parents make copying mistakes in every offspring. Natural selection would be overwhelmed and the species would disappear in a mutational meltdown. What we have not yet done is to ask what relevance this abstract reasoning has for real life, here on Earth. If the error rates of Earthly life are low, well below one per offspring, the theory of the mutational meltdown would have little interest. If, however, mutation rates are high, up near the limit of the tolerable, then that same theory becomes crucially important for understanding life and the human predicament.

So we should turn to the facts. This chapter is a survey of real, live mutation rates, and I have structured it as a narrative from the origin of life to modern human beings. I should say before we set out that we are not off on some well-rehearsed, cruise-controlled tour of definitive, polished science. We are going to deal with frontier science, in something more like an old-time jalopy ride. The breakdowns and backfires will be as memorable as the lurches of forward progress, and just when everything seems to be going right and the romantic vehicle is speeding through virgin territory, it may well blow a gasket and spray the passengers with engine fluid. At the end of it, I could sympathize with someone who reduced it all to a cruel summary sentence – we do not know the harmful mutation rate in any living creature. But that is not the only possible reasonable summary. It could also be said that the subject is being shaken up, and while in former years we relaxed in the belief that mutation rates were too low to pose any problems, now several lines of evidence have grown up, none of them infallible, but all pointing to rates of mutation close to or beyond what in Chapter 3 we deduced to be possible. Human beings, in particular, seem to make about 200 copying mistakes every time a child is produced.

This chapter would then sound like three great trumpet blasts, waking us from our dogmatic slumbers, to the task of explaining how complex life can exist, how natural selection can cope, with this Darwin-defying mutational mess of deadly errors and genetic corruption.

Measure for error

There is an inherent problem in the measurement of copying error rates in the DNA. If we want to know the rate at which human scribes or typists make mistakes, we can simply watch them. We give them a piece of text, have them copy or type it, and count the mistakes. But this method does not work for life because some of the mistakes will have disappeared before we can see them. In the Chinese whispers game, we had a message saying WHEN THE TIGER COMES, FREEZE. We can watch the rate at which new versions of this message arise, but some of the 'mutations' will be invisible because *the tigers have eaten them*. Natural selection continually acts in living species to remove copying errors, and that means we cannot see them all. The observable error rate underestimates the true, underlying rate of error.

Biologists have thought of several ways round, or partly round, this basic problem. One is to make the measurements in conditions in which natural selection is not acting against the mutations. For a message about tigers, we could experimentally keep the tigers away; then the erroneous messages should not be lost and the observed error rate equals the full error rate. A second solution is to look for regions of the message where natural selection probably does not act. Real DNA messages seem to contain regions where mutations are harmless: a change can happen without affecting the creature it is in. It is as if the children's message were WHEN THE TIGER COMES, FREEZE, DE DUM DE DUM DE DUM. They can add or subtract a DE DUM, or change a DUM to a DUN or PUM, without making any difference to their chance of survival. Our measurements of the error rate can be concentrated on the DE DUM region, where the tigers are not acting to remove errors.

The tendency of natural selection to remove errors is only one problem. Another is the difficulty of observing the message itself, and mutant versions of it. We can simply listen to a verbal message, and hear the mistakes; but DNA is not directly accessible to any of our five senses. It has to be isolated and laboriously transcribed by chemical techniques. We shall meet various short-cuts to solve this problem, but they can create problems of their own. A third problem is that we are interested only in harmful errors. Some methods measure the total number of errors, and we then have to estimate the fraction of them that are harmful. Errors can be good, bad or indifferent. We can assume that the good ones are a negligible fraction of the total, but many of the mistakes may be indifferent rather than bad. Sometimes the fraction of errors that are harmful can be conjured out of the total; sometimes it cannot. For all these reasons, the measurement of error is a fallible exercise. Fortunately the theory of Chapter 3 does not require us to be all that exact. It does not matter whether the harmful error rate is 0.7 or 0.75. What matters is whether the rate is well below 1, approximately equal to 1, or unambiguously more than 1. A species with a harmful error rate well below 1 should have no problem in clearing its mutations; if the rate equals 1, the species is near the limit of the possible; if the rate is more than 1, the species poses an existential paradox.

Error rates are most difficult to estimate at the origin of life. We do not have the message; we do not know what language it was written in; and we do not know how it was copied. It is as if the Chinese-whispering children no longer exist for us to observe or experiment on. But even here an uncertain inference can be made. (Table 3 is a summary of the measurements for the whole chapter.)

The primordial soup opera

Life, at its origin, was presumably as simple as it could be: it did not have bodies, with heads and tails, or even any cells, or enzymes; life was just self-replicating molecules. Most life forms on Earth now use DNA as their replicating, hereditary molecule; but they also use the related molecule RNA for several sub-servicing functions. The RNA molecule is capable of replication, and is in several respects a simpler molecule than DNA. For instance, RNA mainly exists as a single strand, unlike the double helix of DNA. A DNA molecule has to be unwound, or 'unzipped' as the geneticists say, before it can be copied or read. Special molecules are needed to unzip the DNA, whereas RNA can be copied or read more directly.

Table 3. Error rates in various forms of life. Figures within a column are not necessarily comparable, because they may have been obtained by different methods (see text). Also, the figures are per generation and the generation lengths are not the same in the different creatures. I have put brackets around figures that are deductions rather than measurements. The length of DNA is the diploid figure, for ease of multiplication for the total mutation rate; biologists may be more familiar with the haploid figure. Numbers of genes, however, are the haploid figures, as in Chapter 2. The percentage of DNA that is coding means the percentage of all the DNA (column 2) that codes for genes (column 5). It is sometimes used for a rough guess at the fraction of all mutations that are harmful.

Species	Mutation rate per letter per generation	Length of DNA (in letters)	Total mutation rate per generation	Harmful mutation rate per generation	Number of genes	Percentage of DNA that is coding
Original life	10^{-2}	[100]	[$<$ or ≈ 1]	[$<$ or ≈ 1]	10–100	100
RNA viruses	10^{-3}–10^{-5}	$< 30{,}000$	$<$ or ≈ 1	$<$ or ≈ 1	100	100
Bacteria (*E. coli*)	5×10^{-10}	4×10^6	2×10^{-3}	2×10^{-4}	4000	100
Worm	$c.\ 10^{-8}$	2×10^8	2	$c.\ 0.5$–1	19,000	30
Fruit fly	$c.\ 10^{-8}$	3.6×10^8	4	$c.\ 0.5$–1	14,000	10–15
Human being	3×10^{-8}	6.6×10^9	200	2–20	60,000	3

For this and other reasons, an 'RNA world' probably preceded the evolution of DNA-based life. The earliest life may not even have used RNA; it may have used some other molecule that had the capacity to replicate. We can, however, reason about copying error rates in a life form that consisted of a molecule like RNA. The molecule consisted of a sequence of letters, such as ... UCGGAC ..., and it copied itself in a soup containing the constituent letters U, C, G and A (RNA differs from DNA in the use of the base U instead of T). The accuracy of the copy depends on the chemical affinities of the replicating molecule, which acts as a substrate, and the resource units in the soup. The molecule might have been like modern RNA in that the letters bonded in complementary pairs: U binds to A, and G binds to C. The sequence UCGGAG then copies itself as AGCCUC, which in turn copies itself as UCGGAG. The accuracy of the copying depends on how likely it is that a C, a G or a U will bind a U rather than the A that is complementary to it. Chemically clued-up experts can estimate this figure, and they have suggested that the error rate would have been 1 in 100 at best. A molecule a hundred letters long would on average copy ninety-nine of them correctly and make one mistake. In the Chinese whispers analogy, this estimate corresponds to the case where we have no children, no message and no game is actually being played. All we can do is consult an expert, who may be able to suggest what the error rate in the game would be, from a general knowledge of human communication.

Manfred Eigen is a Nobel prize-winning chemist who in recent years has turned his attention to the question of how original life, and simple life forms, maintain themselves in the face of copying error. One way that a molecule can do so is by being short. He used the error rate of 1 in 100 to deduce that the original live molecule was less than (or not much more than) a hundred letters long. If the molecule was less than a hundred letters long, natural selection could preserve it. A ten-letter molecule could be preserved indefinitely with ease. There are a million or so (4^{10}) possible sequences of a molecule that is ten letters long, with four different letters. In the local conditions one (or a few) of the million would probably be able to copy themselves faster than the others, and they would eventually take over there. They could also perpetuate themselves indefinitely. The molecule is ten letters long and has an error rate of 1 in 100: about 90 per cent of the offspring molecules will be identical to the parent molecules, an ample supply to take the system forwards.

The form of life, and the way that natural selection would have been shaping it, are so obscure at the primordial stage that we can only guess why complexity might have increased. Darwin thought about the question inconclusively. He once wrote to the geologist Charles Lyell about a question 'which is very difficult to answer, viz. how at first start of life, when there were only simplest organisms, how did any complication of organisms profit them? I can only answer that we have not facts enough to guide any speculation on the subject.' We have more facts now, but they are still inadequate, and Darwin's answer still holds. However, it is easy to believe that there was some advantage. A twenty-letter molecule might hog resources better than a ten-letter molecule, or it might occupy some chemically useful site in the rockpool more successfully. If a twenty-letter molecule did have an advantage, it could evolve. So could a thirty-letter molecule, or one with forty, or ninety, letters. But once we reach a hundred letters we are near the limit of what is possible. Almost every offspring copied from its parent will contain a copying mistake. Either natural selection has to kill every offspring to keep the system pure – in which case it will go extinct – or the molecule's sequence will be progressively randomized over time – in which case it will also go extinct. Natural selection may have favoured increasingly complex life forms, but the mutation rate limited the ability of these complex forms to evolve. A replicating molecule that was 200 letters long might theoretically have been able to out-compete the best 100-letter molecule. That 200-letter molecule could not evolve because copying errors would be made faster than natural selection could remove them: even if the 200-letter molecule somehow came into existence, its sequence would rapidly randomize. A random chain of 200 letters would not be fit to compete against anything.

The argument has assumed that all mistakes make things worse. The figure of 1 in 100 is an estimate of the total rate of copying errors, not the rate of harmful errors. It is possible that none of the mistakes is harmless; 1 in 100 is then the harmful error rate as well as the total error rate. It is also possible that several sequence variants of the original living molecule would work equally well. A mistake that changed the sequence from one equally good version to another equally good version would be harmless (like changing

TIGER to TYGER), a type of change that we can ignore. If only 10 per cent of the mistakes harm the molecule, the other 90 per cent being neutral, the effective error rate shrinks from 1 in 100 to 1 in 1000. The original molecule could then evolve up to 1000 letters or so. We remain so clueless about early life that we have no idea what fraction of changes were harmful. We therefore do not know whether 100, or some higher number, was the upper limit on the length of the molecule. What is compelling in Eigen's argument is not the exact number, but the abstract logic, which suggests that copying accuracy set some limit on the complexity of early life, perhaps in the 100–1000 letter range. The first live molecules might have been less than 100 letters long, and copying accuracy might not then have been limiting complexity. But the limit was probably approached at some stage, because we know that life evolved to be much more complex than a replicating molecule 100, or even 1000, letters long.

> **Mutation storm blows.**
> **RNA polymerase,**
> **Raise my genes again.** *

How did life forms with longer messages manage to evolve? The answer is that mutation rates were driven down below a rate of 1 in 100. The first stage in the reduction was the evolution of copying enzymes. A copying enzyme is a protein that improves the rate and efficiency of the copying process – in DNA or RNA. In a chemical sense, a copying enzyme is a catalyst; it catalyses the copying process. Original life copied itself by the tendency of the complementary bases to line up: a G next to a C, and so on. The simplest imaginable life form with a copying 'enzyme' would be an RNA, or RNA-like, molecule that (amazingly) possessed the ability to catalyse its own replication. Some RNA molecules can catalyse some of the metabolic processes of life. Some RNA molecules can even catalyse the replication of other RNA molecules, but so far no one has discovered an RNA molecule that catalyses the replication of itself; probably it is only a matter of time. Meanwhile, the best we can do to estimate the error rate in simple life forms that have copying enzymes is to turn to that indefensible

---

* As section headings, barbarous biohaiku divide the chapter.

set of creatures, the RNA viruses. The RNA viruses are a molecular chamber of horrors. The rhinoviruses, including the influenza viruses, are RNA viruses; poliovirus is an RNA virus; HIV is an RNA virus. Never has so much suffering been caused by so little nucleic acid. They are among the simplest life forms on Earth, and the simplest for which we have direct measurements of copying error rates. Modern RNA viruses are parasites and depend on more complex creatures; but they may have error rates like those of early life forms. The RNA viruses copy themselves by various enzymatic procedures, but it seems that they all use a raw copying enzyme – a polymerase, in biological terms – and lack the more advanced error-correcting apparatus that we shall meet in more complex creatures. The RNA viruses may therefore reveal the copying accuracy of a life form that has evolved a simple copying enzyme. It is worth noting that a copying enzyme is itself a huge advance on the enzyme-free copying procedures at the origin of life: there were probably many stages in the history of life between original life without copying enzymes and later life with copying enzymes as efficient as those in modern RNA viruses.

The mutation rates of several RNA viruses have been measured. The method used was to measure the rate at which detectable new forms appear in a population of the viruses over time. It is the same method as listening for the children with mutations of WHEN THE TIGER COMES, FREEZE. We saw the problems with the method. It is liable to overlook any errors that are removed by natural selection. It tells us about all errors, not just harmful errors. It also, in this case, only looks at genetic changes that the experimenters happen to be able to detect; these changes may or may not happen at rates that are typical of all the genes in the virus. The method may therefore either over- or underestimate the true, underlying harmful error rate. The experimenters can reduce the problem that some mutations may be removed by natural selection by trying to use conditions in which selection does not act against the mutations that are being counted. They try to keep the tigers out. The second problem, that the harmful mutation rate may be less than the total mutation rate, can be argued against. It is possible that almost all mistakes in RNA viruses are harmful. The code in an RNA virus is very economically written. It has no junk, and contains overlapping genes in which the same bit of code is used in more than one gene. Human DNA contains huge regions that look as if they are of little use, and changes in them may do little harm; but RNA viruses lack such regions – it seems they make every letter count. As we shall see, the various problems more or less cancel out for RNA viruses, and the total rate of observed error may be a reasonable estimate of the underlying rate of harmful error.

The estimates of copying error rates in RNA viruses range from 1 in 1000 to 1 in 100,000 (10^{-3} to 10^{-5}). The range could be real, and some RNA viruses copy their RNA more accurately than others; or it could be noise, and all RNA viruses have similar copying accuracy but different experiments produce a range of estimates. I shall use a figure of 1 in 30,000 as a mid-range estimate of the copying error rate of RNA viruses. The figure implies that the evolution of copying enzymes improved the copying accuracy of life about 300-fold. Life without enzymes has a copying error rate of about 1 in 100 letters; life with copying enzymes has an error rate more like 1 in 30,000. This is an impressive achievement; and as the error rate went down, the complexity of life evolved up. The RNA viruses are indeed more complex than original life. They have several genes – perhaps five to fifty, depending on the virus – coding for several proteins, which are enough to subvert a creature as complex as a human being. An influenza virus, as we all know, is perfectly capable of turning round a human body to fulfil viral, rather than human, goals.

The RNA viruses appear to be operating close to the upper limit of complexity. Real RNA viruses are mainly in the size range of 3000 to 30,000 letters. With a harmful error of 1 in 30,000 letters, they could not evolve to be much longer than they are. We even have some supporting evidence. For a start, RNA viruses are economically organized. They may have to squeeze all the complexity they can into a 30,000-letter molecule, rather than evolving a longer molecule. Also, a remarkably high fraction of RNA viruses are duds, as if they contained mutations that inactivated the virus. For one group of RNA viruses that infect bacteria, only half the viruses manage to smuggle their RNA into the bacterial host. This is feeble, compared with the almost 100 per cent success rate for comparable DNA viruses (DNA viruses have much lower mutation rates than RNA viruses). The RNA viruses may only be able to survive by running off huge numbers of offspring copies from each parental virus. If they try often enough, then even with their erratic copying they may manage to run off a perfect copy every now and then. One RNA virus is known to make over 1000 offspring copies from each parent.

The fraction of functioning viruses can be brought down further by experimentally increasing the mutation rate. One group of virologists, led by John Holland, used chemical mutagens to triple the mutation rates of two RNA viruses. The result was a complete loss of infectivity, presumably because all the copies of the viruses were now riddled with errors: the viruses had experienced an experimental mutational meltdown. We can now see why the 1 in 30,000 figure may be a reasonable estimate of the harmful error rate, despite the problems in the method used to obtain it. If the real rate of harmful error were much less, it would be puzzling that a mere three fold increase could tip the viruses into a meltdown. If the real harmful error rate were much more, it would be puzzling that RNA viruses exist. We can also tentatively conclude that some RNA viruses have evolved near to the limit of complexity set by their copying error rates. Original life may have been a molecule about a hundred letters long with a copying error rate of about 1 in 100. Its complexity was limited by its error rate. Then life evolved the use of copying enzymes, reducing its error rate to more like 1 in 30,000. The complexity of life increased by the same factor, about 300-fold, to a molecule about 30,000 letters long. The complexity of life continued to bump up against the limit set by its error rate.

I like Holland's experiment because it points to what is currently my second favourite hypothetical cure for AIDS. We just need to increase the viral mutation rate and they'll collapse under their own copying errors. The treatment would need selectively to increase the viral, and not the host's, mutation rate, but that should be feasible. The virus would probably respond by evolving to copy itself more accurately; strains of HIV are known that have lower error rates. But even that may not help them, because the improved copying accuracy is bought at the price of slower reproduction. Our immune systems should be able to fight better against a slow-breeder virus. There would be a certain beauty in this cure. I often see it said that the high mutation rates of HIV give it an advantage against its human hosts, allegedly because it allows them to evolve faster against us. On the argument of this book, however, mutations are a net disadvantage; more means worse, when we are talking about mutation. I should like it if mutations turned out to be the Achilles heel of HIV. If that sounds too much like wishful thinking, the argument so far does offer another reason for hope. The RNA viruses may not be able to evolve much of an increase in complexity; they have to fight us within their existing informational limit. They will probably spring many new tricks on us as they are, but it is reassuring that one class of extra tricks is denied them.

> **Deep inside me now,**
> *Escherichia coli*
> **Copies genes with care.**

Our next stop is to visit the bacteria. There have been two developments by this stage: one is that bacteria carry their genes in the double-stranded (that is, double helix) molecule DNA, rather than single-stranded RNA; the other is that bacteria have an apparatus of about fifty proofreading and repair enzymes. The result is that they copy themselves much more accurately. The copying error rate per letter in bacteria is about 1 in 1000 million, when measured by the same method as gave the figures of 1 in 1000–100,000 for RNA viruses. The estimates, as we saw, are uncertain, but this may not spoil a comparison between the two kinds of creature. Bacteria seem to make copying mistakes at a rate 10,000–1,000,000 times lower than in an RNA virus. If we use the mid-range figure, the evolution of DNA and the error-correcting enzyme machinery combined to reduce the rate of copying error 100,000-fold.

When copying enzymes reduced the error rate 300-fold, the complexity of life increased 300-fold. But when the next set of improvements reduced the error rate a further 100,000-fold, life did not immediately exhaust the new opportunities. Bacteria are only about a hundred times more complex than RNA viruses. The bacterium *E. coli* is fairly large as bacteria go. It has four thousand and a bit genes and 4 million letters of DNA code. We can calculate how many mistakes *E. coli* makes when it breeds. Its error rate per letter is about 10^{-9} to 10^{-10}; let us use 5×10^{-10}, a number halfway through the range. The approximate total number of errors when *E. coli* breeds is then $4 \times 5 \times 10^6 \times 10^{-10}$, or 0.002. This number is much, much less than 1. Of every 500 baby *E. coli*, 499 are identical to their parent. No matter what fraction of the mutations do harm, natural selection will have no difficulty eliminating them and maintaining the quality of the bacteria. Until the evolution of bacteria, DNA messages had been as long as they could be, given their rate of copying error. Then the error rate was reduced by a whopping five orders of magnitude. Bacteria are more complex than their predecessors, but by only about two orders of magnitude. During bacterial evolution, something other than copying error has set the ceiling on complexity.

> Proofreading enzyme
> Carry my whispered message
> Past the tiger's teeth.

The 100,000-fold reduction in error rates is quite some improvement, and it is worth looking at how it has been achieved. I have tended to use the terms 'mutation' and 'copying error' as if they were almost interchangeable, as if all mutations happened when DNA is copied. Many indeed do happen then, but they can also happen by spontaneous chemical change in the DNA, or the RNA, at times other than when it is being copied. Like all large biological molecules, DNA and RNA tend spontaneously to fall to bits. Both molecules have a structure consisting of a backbone made up of a long chain of alternating phosphate and sugar molecules. The sugar is called ribose in RNA; in DNA it loses an oxygen atom and is called deoxyribose. The information is coded in a third component, called a base, which is attached to the sugar ('base' is a technical term in chemistry, unrelated to its ordinary language meanings to do with wickedness or structural support). The letters A, G, C and T in DNA and A, G, C and U in RNA refer to different bases. A nucleotide consists of a unit made up of one base, one sugar and one phosphate. As we saw earlier, DNA is double-stranded, with two strands of nucleotides arranged complementarily in a double helix; RNA is generally single-stranded.

It turns out that DNA is a more stable molecule, much less vulnerable to spontaneous change than RNA. One kind of spontaneous mutation happens when a base drops off its sugar, leaving an unadorned backbone of sugar and phosphate. This happens at a higher rate in RNA than DNA; indeed, deoxyribose is chemically peculiar in its tenacity for bases. But even in DNA bases do spontaneously drop off. One expert, Tomas Lindahl, calculated (using measured rates for test-tube DNA) that every day in every human cell 2000–10,000 bases fall off their sugars in the DNA.

In us, this is of merely chemical interest because we have special repair enzymes that put the bases back again. Bases that fall off the helix are just one way the DNA degrades. Bases also spontaneously change into other bases. Base C is the main culprit: it tends to lose part of its structure, converting itself into another base called U. Conversion of C to U is so common that we have an enzyme that does nothing except correct these mistakes. Another snag is that neighbouring bases in a strand may form cross-bonds called dimers; this kind of mutation is particularly caused by ultraviolet light (the mutagen of sunlight – these dimers are part of the story of skin cancer) and there are yet more enzymes that detect and remove dimers.

The double-strandedness of DNA also makes it more stable than the single-stranded RNA. Bases fall off single-stranded DNA at four times the rate of a double-stranded DNA. Bases are also more likely to be spontaneously transformed in single-stranded DNA. The base C transforms into U in single-stranded DNA at 200 times the rate of double-stranded DNA. The RNA molecule is single-stranded and therefore has a higher rate for these mutations than double-stranded DNA. Moreover, as we saw in Chapter 3, the redundancy inherent in the two strands of the DNA makes it possible to repair some faults. In all, DNA is inherently more stable than RNA, and its double strands also allow superior repair after damage or spontaneous decay.

There is another contributor to the 100,000-fold reduction in error rates between RNA viruses and bacteria: it is that RNA viruses lack proofreading enzymes in their copying process. Proofreading works as follows. When the DNA is copied, the two strands are unzipped and the complementary sequence is copied on to each strand. If the strand reads … CATTACA … it will have … GTAATGT … copied next to it. A copy mistake may be made, such as a T instead of a C. It will create a mismatched T–G pair between the copy and the original strand. In normal DNA, we should know that a T–G pair was wrong, but not which letter of the pair was wrong. In copying, however, it is possible to tell because the original and the copy lie side by side. If you are supervising a scriptorium of monks, you can look over their shoulders and see the original next to the incomplete copy. If they differ, it is the new copy that needs correction. In the same way, the DNA copying machinery 'knows' which is the original strand and which is the copied strand. If there is a mismatch, it is more likely to be the copy, than the original, that is wrong. The special proofreading enzymes detect mismatches as they arise, and has the letters recopied.

The full set of repair and proofreading enzymes no doubt evolved in many stages. Error rates then also improved in many stages. But we lack any evidence of what those stages were, and can only see their final, combined result: the splendid copying that we can read about for bacteria in Table 3. Proofreading and repair enzymes (and DNA) are all absent from RNA viruses, all present in bacteria, and much the same set of enzymes are used in bacteria and all the rest of life, including us. The combined effect of the enzymes plus DNA was to cut the error rate by a factor of about 10^5, down to 10^{-9} or 10^{-10} per letter. That figure then appears to be approximately constant in bacteria and in more complex life including us, implying that the copying accuracy per letter is the same in all life except viruses. This suggests that the multienzyme, error-reducing apparatus that we first see in bacteria operates best as a package. It was valuable to acquire, but once acquired was prohibitively expensive to improve; the error rate reduction was a one-off event.

In Table 3, the number for the human error rate per letter may appear to differ from the number for bacteria. But the two numbers are not simply comparable. I have written the numbers per organismic generation, not per round of DNA copying. Bacterial DNA is copied once when a new bacterial cell is reproduced; but there are many cellular generations between conception (as a fertilized egg) and reproduction in multicellular creatures like us – an estimated 33 divisions in women, and about 200–600 divisions in men. If we allow for this difference, the rates of error per copying event per letter look about the same in bacteria and humans. Also, the bacterial and human figures in Table 3 were obtained by different methods, as we shall see when we reach human beings later in this chapter. It is uncertain that the error rates per letter really are similar in bacteria and in more complex life, and some life forms may have error rates above or below those in Table 3. But it is a good generalization, with suitably magnificent gestures of uncertainty, that the copying accuracy of a letter of DNA has been roughly constant for all cellular life for the past 3500 million years.

And the word was made flesh

The evolution of DNA, and the error-fixing enzymes, put the Chinese whispers problem into evolutionary suspense. There was plenty of scope for life forms far more complex than bacteria to evolve. After bacteria, the eukaryotic cell was the next big step in the history of complex life. Early eukaryotes were single-celled creatures, just as bacteria are single-celled, but the single cell of a eukaryote is more complex than a bacterial cell, and is coded for by more genes. However, the error rate had been driven down so far that no new error-fighting devices were needed at this stage. The early single-celled eukaryotes continued to produce an ample supply of offspring without any copying errors. For example, the common yeast that is used by bakers and brewers is a kind of single-celled eukaryote. Its DNA is about 12 million letters long, and contains about 6000 genes. Its mutation rate has been measured by the same method as was used in RNA viruses and the bacterium *E. coli*. It turns out that the copying error rate per letter is much the same in yeast and bacteria, about 1 mistake per 1000 million or per 10,000 million letters (10^{-9} to 10^{-10}). A yeast cell therefore makes about three times as many mistakes when it breeds as does an *E. coli* cell, because its DNA is three times as long, but its mutation rate is still tiny. Mutation poses little problem for the evolutionary persistence of yeast.

Time passed, and some single-celled eukaryotic creatures started to evolve into multicelled creatures. In the simplest multicellular forms of life, all the cells are much the same; but it was then only a small step to the evolution of egg-to-adult development. An organism with development starts life as a fertilized egg, and that cell then divides in two. The resulting cells divide in turn, and as the cells successively divide they start to develop into particular kinds of cells: some cells develop into muscle cells, others develop into brain cells, and so on. The evolution of development was clearly a great event in the rise of complex life; a body with development is a more complex form of life than an undifferentiated mass of cells or a single-celled life form. What matters about development here is the way it influences mutational error. It has two direct influences, and seems to be associated with a third, indirect influence.

One direct influence of development on error rates is to reduce them, or at least to reduce their effects. The developmental process contains mechanisms to put things right again after accidents. Accidents happen for various reasons, and mistakes in the coding instructions are one of them. The developmental accident-fixing mechanisms cannot correct errors in the DNA code, but they can prevent the errors from doing damage in the body. They disguise, rather than correct, error. An instruction like 'when the tiger comes, freeze' is an inadequate analogy here. The instruction directly leads to the action; there is no developmental process in between. The genetic instructions that underlie development cause the form of a body to change over time, and are more analogous to the instructions in a cooking recipe. Recipes contain instructions like 'heat the ingredients, stirring continually, until the sauce thickens'. The result is that the raw ingredients are transformed into something else – sauce. But accidents happen in cooking, and a recipe may also include troubleshooting clauses. If you heat too hard, or stir too little, or become distracted and go away, you will need an instruction like 'if the sauce shows any tendency to be lumpy, either beat it with a small wire whisk or put it in a blender for a few seconds'. Some accidents cannot be fixed in this way; food that is burned is permanently spoiled. But other accidents can be fixed, and then a troubleshooting instruction is worth having.

The development of living creatures can, like cooking, go wrong. The developing organism may suffer some shock such as poisoning or starvation; or some cells may misinterpret a signal and grow in the wrong place. The developmental programs of our bodies contain instructions that can rectify some of these temporary misfortunes. What matters here is that a genetic accident is an accident like any other. The developmental troubleshooting mechanisms can work against genetic as well as non-genetic mistakes. A remarkable experiment in 1998 illustrated the point. The experiment used fruit flies. One of the genes that helps to fix mistakes in fruit flies, and in other life forms including us, is the one coding for a molecule called heat shock protein 90. This protein helps cells to protect themselves against many kinds of stress, including the heat shocks it is named after. In the experiment, the protein was experimentally inactivated or distracted during the development of some fruit flies; the result was that many of the flies grew up as monsters. Some had deformed eyes, some had extra antennae, some had wing-edge material growing in the middle of the wing, one had a 'sex comb' growing on a 'transformed second leg'. The list goes on. The reason, or part of the reason, why they grew up as monsters is that every fruit fly (like every human being) contains genetic defects. Normally these defects are put right during development by genes such as heat shock protein 90. But if the troubleshooting genes are distracted, the defects are no longer fixed so well, and the full range of underlying error is brought to light. In terms of the cookery recipe, it is as if you 'stress' the cook, by a distraction, or by filling the blender with mud. The result will be spoiled cooking. In a more exact analogy, it is as if the recipe itself contains erroneous instructions as well as troubleshooting clauses. If you block the equipment that is used in troubleshooting, the effect of the instruction errors will show through. What the experimenters did was to sabotage the blender.

The gene producing heat shock protein 90 (*hsp 90*) is one example of a troubleshooting gene. Another is a gene called *p53*. Gene *p53* is famous because it protects us from cancer: *p53* is defective in 80 per cent of human cancers. A proper, non-defective *p53* gene is activated when the cell is damaged, either by external mutagens such as ultraviolet radiation or by internal horrors such as spontaneous decay and copying error. The *p53* gene then orders the cell either to stop dividing or to commit suicide. In the gastronomic analogy, *p53* does the equivalent of cutting out a small, damaged part of an otherwise satisfactory dish. It cannot correct the recipe that led to the damage, but it can recognize something that is inedible and cause it to be excised. It is not a repair enzyme, but it can prevent mutated DNA from damaging the body, by killing off the cells that contain the mutated DNA. (It has recently been discovered that *p53* can also order DNA repair, though it does not do the repair; it is probably part of a general damage-response team. It surveys damaged DNA, and can try a range of responses, from local repair, to putting the cell in suspended animation, to killing the cell, depending on the scale of the damage.)

Troubleshooting works rather like health care. If someone contains a defect (such as a genetic defect), a doctor may be able to improve the quality of the individual's life by prescribing some appropriate exercise regimen or diet, or even by surgery. The doctor knows how to fix the fault despite not knowing how to correct the DNA error that produced the fault. Our bodies also contain internal mechanisms that know how to fix faults but do not know how to correct DNA. Why is it that live systems can 'know' how to make up for errors in their instructions, without knowing how to correct the DNA code? Living creatures can correct copying errors in the DNA only at the time the errors are made, when the original strand lies next to the copied strand and the two can be compared. The developmental mechanisms operate several stages downstream. A fly's developmental system 'knows' how many antennae there should be and can override an error in the instructions that contribute to antennal development. But these mechanisms cannot go back and correct the error in the DNA, probably because the reverse translation problem is too difficult. By reverse translation, I mean that the body would need a mechanism that can work back from the form of the body to correct an error in the digital code. This is not easy. There are likely to be many possible causes for any one accident. It could be difficult to work out what the cause was, whether it was genetic at all, and if so which gene contained the error. Even if you identify a gene that contains an error, there is still the problem of working out what changes are needed in the code. Scientists are unable to perform the reverse translation, from body to DNA code, except in one or two easy cases. It seems that biological mechanisms have been unable to do it either.

Living systems have proofreading and repair enzymes that correct some genetic errors, but not all. They also have internal-care mechanisms that fix a further fraction of the errors that have made it through the proofreading and repair enzyme filters. Some internal-care mechanisms troubleshoot the molecular processes that operate inside each cell, and single-celled creatures have these mechanisms, just as we do. Other internal-care mechanisms are only useful in many-celled creatures. The action of the $p53$ gene to make a cell with damaged DNA commit suicide, for example, would not be advantageous in a single-celled bacterium. Cell suicide is an error-disguising strategy that arose only after the evolution of multicellular life. All the troubleshooting mechanisms act to neutralize mistakes: they make a harmful error harmless. One influence of the evolution of development, therefore, was to reduce the fraction of mistakes that cause damage.

The other direct influence works in the opposite direction. The evolution of development also saw an increase in the error rate. The reason is that the DNA has to be copied many times before an organism reproduces. The organism starts life as a single cell, and this cell successively divides over time. One subline of cells is destined to form the reproductive cells: the cells that will become sperm or eggs. The DNA has to be copied once for every cell division, and there are several cell divisions per life cycle. The number of cell divisions depends on how long the creature lives, and on how many sperm or eggs it produces, but the general trend is clear. Development increased the number of times the DNA was copied between reproductive events, and more complex creatures have longer developmental cycles and more cell divisions in a lifetime than simpler creatures. If the basic error rate is 1 in 1000 million letters per copying event, then an organism that goes through 100 cell divisions before breeding will see its error rate increased 100-fold, to about 1 in 10 million letters, by the time it breeds.

A third influence on mutational error also came into play at about the same time as the evolution of development. Between bacteria and multicellular animals, we start to see evolutionary increases in the number of copies of the gene sets in an organism – in its 'ploidy'. We saw in Chapter 3 how errors in one gene set can be masked by a second gene set. There is a crude trend, in which simple life forms usually contain a single gene set and complex life contains two gene sets. Most bacteria are haploid, whereas most multicellular animals, including us, are diploid.

The evolutionary story about multiple gene sets is not straightforward, however. Between the haploid bacteria and the diploid animals there are all sorts of arrangements in plants and in microbes. A ciliate such as *Paramecium* is a single-celled eukaryote; it is more complex than a bacterium. *Paramecium* has its DNA in two nuclei, and in one of them the gene sets are multiplied up to hundreds of copies. Other microbes have their own polyploidal variations, and plants also have a much more liberated attitude to gene numbers than do bacteria and animals. No one has ever made sense of these details, but the broad trend looks right. As gene numbers increased from bacterial levels, life probably had a problem in coping with mutations at some point; and it is in creatures with more genes than bacteria that we see the evolution of multiple gene sets. In animal evolution, the number of gene sets per individual soon settled down to the diploid arrangement that we use. The evolutionary changes in ploidy seem to happen roughly when we might expect in the history of live complexity.

Error rates were cut to such a low level in bacteria that copying error was no problem during the evolution of the eukaryotic cell. Yeast are single-celled eukaryotes and, as we saw, have negligible error rates. As multicellular life evolved, with several cellular generations in a life cycle and with expanding numbers of genes, the total error rate would have started to climb. But did it merely climb from one negligible number to another, slightly higher but still negligible number? Or did it climb all the way back up to the ceiling that life in the RNA world was squeezed against?

Ripening vineyard
where the sun-kissed fruit flies play
Mukai tests their genes

Our next stop is with fruit flies. Fruit flies use about 14,000 genes, five to ten times more than bacteria. They are unambiguously multicelled and have egg-to-adult development. A fruit fly starts life as an egg, develops into a worm-like stage, and later metamorphoses into an adult fly. There are about a thousand species of fruit flies, which live off many kinds of fruit; biologists happen to have concentrated their researches on a species that lives around vineyards. Fruit flies are the appropriate place to introduce a superior, or potentially superior, method of estimating harmful mutation rates. The method was invented by a Japanese geneticist, Terumi Mukai, in the 1960s. The method does the equivalent of keeping the tigers away from a houseful of Chinese-whispering children. Their message then deteriorates because there are no tigers to remove mutant messages. Every ten generations or so, Mukai did the equivalent of taking a sample of the children and putting them in the tiger cage. He then measured their rate of survival in the cage. As errors accumulate in the tiger-free lines of children, the message will come to provide less and less accurate instructions about how to deal with tigers. The sample children's survival in the tiger cage will accordingly go down between successive samples, and the rate at which it goes down can be used to measure the rate at which errors are happening in the protected line. In practice, Mukai kept lines of fruit flies protected from selection, and then sampled them from time to time and measured their viability. Viability was measured as the fraction of eggs that grew up to be adults. The viability of the sample flies should decline between successive samples as mutations accumulate. The experiments are hard work, because many lines of flies are needed to provide measurements. Mukai and his assistants had to rear and check several million fruit flies.

The method has a beautiful logic, but it is by no means easy to implement. If a message simply says WHEN THE TIGER COMES, FREEZE, you can stop selection from preserving it by keeping the tigers out. With real fruit flies, however, the DNA message contains 14,000 instructions about every detail of their lives, and it is much harder, or even impossible, to stop every force of selection. You can stop selection against errors in the anti-predator instructions by keeping predators away. But you cannot stop selection against errors in the feeding instructions and the breathing instructions by making it unnecessary for the flies to feed and breathe. In more abstract terms, natural selection can operate in three ways as one generation gives rise to the next. First, by differences in survival: a bad mutation may reduce its bearer's viability at any stage from egg to adult, and natural selection removes the mutation when the flies that contain it die. Second, by differences in fertility: bad mutations may interfere with egg (or sperm) manufacture. Third, bad mutations may reduce a fly's (usually a male fly's) sex appeal or power in the mating market. Mukai's aim was to put a stop to all three processes, by appropriate experimental tricks.

Selection at the first stage works by premature death, and to stop it we have to put a stop to premature death. We do so in much the same way as we try to prevent death in human beings. Mukai's procedure was to keep the flies in optimal conditions – optimal, that is, for fruit flies: plenty of food, no birds or other predators of flies, no overcrowding, excellent standards of hygiene, the temperature and other environmental conditions all just right. In the ideal case, the experimenters should make the conditions so perfect that all the fruit flies survive: there is no death at all before the stage of reproduction. No one has quite achieved this ideal, but they have come close. For instance, Svetlana Shabalina, Lev Yampolsky and Alexey Kondrashov repeated Mukai's basic experiment, with a few improvements, at Cornell University, New York. They counted the number of eggs each generation, and the number of adults that grew up; 85 per cent of the eggs grew up as adult flies. It may be difficult to improve on this figure, because a certain fraction of eggs are infertile, or unfertilized, or contain gross defects; those eggs will fail no matter how much they are molly-coddled. It may be, therefore, that this experiment reached the ideal of stopping all selection by differences in survival; at any rate, it came close to it.

The second two kinds of selection can be stopped with certainty. The trick is to pick one pair of flies (one male and one female) at random to form the next generation. It is important not to allow the population of flies to breed of its own accord: flies with poor, mutant genes would probably be discriminated against as mating partners, and even if they did pair up successfully they would lay fewer eggs. Males with bad genes tend to produce defective sperm, and females to lay fewer and inferior eggs. Thus if the population is left to breed by itself, natural selection will act against bad genes. We can put a stop to it by forming the next generation from one randomly picked pair. Now the flies have no opportunity to choose partners with good genes, and no fertility differences exist: there is only one pair to do the breeding.

So we have a population of flies, protected from natural selec-tion, and their quality decreases over time. Every few generations we can extract a few flies to see how decent a set of flies they are. Mukai found that the average viability of the flies decreased 1–2 per cent per generation. The observable decrease in viability is a compound effect: it is due both to the number of harmful mutations (which we are interested in) and to the amount of harm that an average mutation does (which we are not). An observed decline may be due to a small number of catastrophic mutations or a larger number of minor mutations, or some mix of the two. The number of mutations has to be statistically extricated from the compound effect. The detailed calculations are wet-towel-around-the-head stuff, and we can ignore them. But it does matter that there is an element of inference in Mukai's method; he does not measure mutation rates directly. When the statistical extraction has been performed, it turns out that a decrease in viability of 1–2 per cent per generation implies that a fruit fly makes about one harmful copying mistake per offspring. Because of the element of inference, it is also compatible with mutation rates below 1 or above 1. Mukai's experiment suggests that fruit flies are tan-talizingly close to the limit of the possible.

Mukai did his original experiment in the 1960s but his work did not attract much interest at the time. In recent years, there has been a great renaissance of interest and many groups have sought to repeat the basic experiment and to improve its implementation. Some of the results suggest high mutation rates, like Mukai's estimate. Others suggest that fruit flies have a negligibly low mutation rate, like those we saw in bacteria. I have watched this stream of research with interest. I have been struck by how, as each seemingly convincing experimental result has been published, you just have to wait a month or two for it to be put in doubt. One result had to be retracted after it became clear that the experimental flies had become contaminated with a kind of genetic disease. Another set of experiments measured the viabilities in conditions that were too 'soft', as if the tigers were all old and fat. I do not think there are any results that everyone agrees are bug-free.

The problem is not that the experimenters are clumsy, or stupid, or that some of them have cooked the books; it is that the experiments are genuinely hard to do. But the ambition is so grand, the experiments are so important, that people will persist with them and the self-correcting nature of repeated science should soon assert itself. I expect it will soon be clear how many harmful mutations happen when a fruit fly, or comparable creature, breeds in a Mukai-type experiment. If forced, I should guess that the high figures, of one or so harmful mutations per offspring fly, will turn out to be right. The experiments that look most reliable point to this kind of mutation rate. The figure also fits in with an independent estimate that can be made for worms and flies by a similar procedure, as we shall use in the next section for humans (Table 3). If the figure is right, then at some evolutionary stage between bacteria and creatures as complex as fruit flies, error rates were pushing back against the upper limit. But I also sympathize with the fence-sitters who insist that the whole field is currently too confused for any conclusion to be drawn.

To err 200 times is human

The harmful error rate has not been measured in human beings, by Mukai's or any other method. But we do have a method of estimating the total error rate. It uses regions of the DNA that seem to be wholly useless. They are the equivalent of the DE DUM DE DUM nonsense in a message that reads WHEN THE

TIGER COMES, FREEZE DE DUM DE DUM DE DUM. The DE DUM part provides no useful instructions, and has no influence on the life of its bearer apart from the time it takes to copy it. Mutations will then accumulate in the DE DUM nonsense at the same rate as they arise. Our DNA contains stretches of code called pseudogenes. Pseudogenes are probably accidental duplicates of real genes that have been made redundant. The DNA in a pseudogene closely resembles a real gene, but lacks some crucial element that all real genes need to function. A pseudogene is a stretch of useless DNA.

It is the uselessness of pseudogenes in the life of their bearers that makes them useful in our argument. Natural selection does not act against any of the mutations in a pseudogene, because mutations in pseudogenes make no difference to the success of the organism. Mutations in pseudogenes accumulate over evolutionary time at a rate equal to the mutation rate. What we need to do is find a particular pseudogene in two species, such as chimps and humans. We then count the number of differences between the versions of the pseudogene in the two species. We know that the difference between the two pseudogenes evolved in the 5–8 million years since the common ancestor of humans and chimps. We can calculate the rate of evolution. For a typical pseudogene, the rate of evolution turns out to be about one change per DNA letter per thousand million years, or one thousand-millionth (10^{-9}) of a change per year. We can multiply this by the human generation length (30 years or so) and the total number of letters in human DNA (6.6×10^9) to find the total number of mutations. The result is about 200 mutations per child. You can play with the numbers, but the result always comes out as a few hundred.

Two hundred or more mutations per child. This is a decidedly big number, and in itself it is a striking fact about human beings. But it is not exactly what we need. It is the total number of mutations, whereas we need the total number of *harmful* mutations. We need to know what fraction of the 200 are harmful. Unfortunately we do not know, but we can reason about it in two ways. We can loosely say that even if only a tiny fraction (such as 1 per cent) of mutations are harmful, human beings will still suffer more than one harmful mutation per offspring. Alternatively, we can use various more concrete arguments to try to extract the number of harmful mutations in the 200 total. Even the most

rigorous accounting has not pushed the number of harmful mutations below 1. The lowest estimates are about 2. Other estimates
are as high as 20. A figure of 5–10 is therefore quite plausible.

Whether the number is 2, 5, 10 or 20, human beings are in the
paradoxical zone. Our mutation rate is higher than a living system
can have, according to the argument of Chapter 3. Natural selection should have been overwhelmed by human copying error, and
human DNA should have been randomized out of existence. Our
mutation rates are so high because we are such complex life forms.
We have long DNA molecules, large numbers of genes, and long
lives during which our DNA is copied many times. An evolutionary theorist who was dragged out of the ivory tower and
confronted with the evidence of human existence might well
remark, after a moment's reflection: 'they are all very well in fact,
but they'll never work in theory.'

That concludes my survey of mutation rates. The important
results are quite recent, and uncertain. Indeed it is because they
are at – some would say beyond – the frontiers of science that they
are uncertain. It is risky to generalize about the opinion of the
scientific community, but until about ten years ago most biologists
probably thought that mutation rates were well below 1 per
offspring in all life forms. Two great biologists from the mid-
twentieth century, H. J. Muller and J. B. S. Haldane, were experts
on mutation rates. They thought that mutation rates were well
below 1, partly because of the evidence at the time but mainly
because of the reasoning (if in more dignified form) of Chapter 3.
Living things just *had* to have mutation rates well below 1. But
the weight of evidence is tipping against the old view. The subject
is controversial, but Eigen has persuaded many biologists that
error was a limiting factor at the origin of life. Research on RNA
viruses, stimulated by the AIDS plague, has revealed that RNA
viruses copy themselves in a slap-happy way and they appear to
live on the edge of a mutational meltdown. Mukai-cossetted flies
seem to deteriorate in quality by 1–2 per cent per generation.
Mukai began his work in the 1960s, but it was only in the 1990s
that his experiments were repeated and some progress made in
debugging his amazing experimental design. Pseudogenes were not
discovered until 1979 and a decade passed before there was enough
evidence to measure their evolutionary rate. The first estimates
inevitably were ropy and people suspended judgement, but later

work has confirmed initial impressions and we can now be reasonably sure that pseudogenes are almost out of evolutionary control. They tick over like crazy. Geneticists now generally accept that 200 or so mutations happen when a human being breeds. The late 1990s saw at least three lines of evidence converging on a mutation rate in many life forms, and in particular in human beings, close to or even above the paradoxical figure of 1 per offspring.

Imagining natural selection

These high error rates tell us something about what natural selection is doing in the real world. Natural selection is famous as an agent of evolutionary change, such as in bacteria that evolve to resist (or even live off) antibiotics and moths that evolve camouflage. Natural selection is also famous as a creative process that builds up complex adaptations, such as the eye and other marvels of engineering design, over longer periods of time. But if the error rate is near 1 then only a small fraction of natural selection is concerned with these processes. The evolution of antibiotic resistance, or dark coloration in the peppered moth, are extreme examples of short-term change powered by natural selection. The selection pressures at work are much higher than average, and much higher than during the evolution of something like an eye from a simpler light-sensitive neuron.

Human beings have imposed huge selection pressures. Take the example of the peppered moth, for which some approximate numbers are known. In Britain there are two main forms of the moth, a light form that is camouflaged in unpolluted areas and a dark form that is camouflaged in places that are polluted by industrial smoke. Measurements indicate that in a polluted area the light-coloured form has a survival rate about half that of the dark form. We can use the numbers to work out the chance that, on average, a moth would have died because it was the wrong colour. Immediately following the industrial revolution, the dark form was still rare and most peppered moths were light in colour: the average moth was then almost certainly the wrong colour and it had about a 50 per cent chance of dying prematurely because of poor camouflage. This was a massive selection pressure, and soon led to an increase in the frequency of dark moths. After twenty or thirty years the dark form made up about half the peppered moth population in industrial regions of England. An average moth was now more likely to have the right colour pattern, and its chance of dying of poor camouflage had reduced to about 25 per cent. As the dark forms increased to a frequency near 100 per cent, the mortality due to poor camouflage declined to a negligible level.

In an extreme case of evolutionary change driven by natural selection, half the moths could have been killed by the process driving the change. Compare that with the death toll from harmful mutation. The mutation rate in moths has not been measured, but is probably similar to fruit flies at about 1 per offspring. About 70 per cent of an average parent's offspring will contain copying errors. These copying errors have to be purged, and the average newborn mothling therefore has about a 70 per cent chance of dying prematurely because it contains a mutation. The death toll due to mutation is a permanent condition, unlike the few years of high selective mortalities of the nineteenth-century peppered moths. Every generation, in a normal moth, natural selection executes more mothlings for reasons of mutational error than it did in the freakish conditions of one species, the peppered moth, while it was driving an extraordinarily rapid bout of evolutionary change. (The situation is probably different in bacteria. Bacteria have lower rates of harmful mutation than moths, and my argument that natural selection is more of an agent to preserve, than alter, life has less force for bacteria.)

Most species, even of multicellular life, are not like the peppered moth in the industrial revolution. No one knows how much directional evolutionary change is going on in a more typical species, but the amount is probably somewhere between not much and zero: only a few per cent of individuals would be dying for reasons of directional evolutionary change. Natural selection would then be killing ten to a hundred times more individuals because of their harmful mutations than it was killing in the cause of evolutionary change. Most of the work of natural selection would be concerned with preserving messages, not adjusting messages to environmental change. Natural selection is described as a theory of evolution, and indeed it is one, but it is a theory of non-evolution too, of how life maintains itself against spontaneous decay. Conservatives will point a political moral about the relative effort we should allocate to preserving civilization, and to adjusting it; radicals will point a moral about the dangers that lie in analogies. But the harmful mutation rate estimates of this chapter support a concept of natural selection that has some claim on everyone's attention.

A syllabus of errors

More work on mutation rates is needed, but I think we know enough to risk a narrative history of error. The narrative will need to be modified as new facts come in, but here is how it looks to me at present. The simplest life forms for which we have evidence are RNA viruses, but there was presumably a long period of evolution in the replicating molecular apparatus before anything evolved that was as complex as an RNA virus. The earliest replicating molecules would have lacked enzymes to catalyse their replication. The error rate was then about 1 in 100, limiting life to a message about 100 letters long – or more if some of the errors were harmless. The first breakthrough for which we have evidence was the evolution of a copying enzyme. If RNA viruses are any guide, the enzyme drove down the error rate by somewhere between 100-fold and 1000-fold. Life could evolve greater complexity, and replicating molecules evolved that coded for several genes. But the complexity of life in the early 'RNA world' was still limited to about 30,000 letters by the accuracy of its replication.

Natural selection then favoured a whole series of genes for proofreading and repair of the replicating molecule. It also favoured the stabler, double-helical DNA over the fragile single-stranded RNA. The next stage for which we have evidence is the bacteria, and even the simplest bacteria have an elaborate set of proofreading and repair enzymes. The enzymes together with the use of DNA cut the error rate to about a millionth of its previous level. Bacteria with DNA molecules a few million letters long exist well below the mutational danger zone. The fossil record shows that bacteria had evolved by about 3500 million years ago, and then there was a long delay before life evolved greater complexity. If we assume that the first bacteria had proofreading enzymes like modern bacteria, it was not the error rate that was causing the delay. Since the evolution of the bacteria, the copying error rate has remained roughly constant at 10^{-10} or so. We use a similar set of enzymes, and the same hereditary molecule (DNA) as bacteria do. But while the copying error rate per letter has remained constant, the length of the DNA molecule has expanded about 1000-fold from the few million letters of bacteria to the few thousand million letters of ourselves. The number of cell divisions per generation has also increased from one in the life cycle of a bacterium to about a hundred in us. The combined effect is a 100,000-fold increase in the number of errors per average offspring, from a trivial level up to about 200 per human child. About two to twenty of those 200 are harmful.

The mutation-fixing skills of the 3000-year-old proofreading enzymes were eventually stretched to the limit. Somewhere between bacteria and ourselves, life bumped back up against the error limit that it had encountered before the evolution of all those clever proofreading enzymes. I do not know exactly when it was. Maybe it was by the time we had multicellular animals like worms and flies – creatures like that had probably evolved by 500–1000 million years ago. I suspect that by then parents were making mistakes in almost all their offspring and the existence of life had become paradoxical, but the evidence is unconvincing. By the time we reach human beings, with their 60,000 genes and 6600-million-letter supercoiled DNA molecules, the paradox can no longer be denied. We blunderingly crank out our offspring with over 200 mutational errors. We should have been randomized in a mutational meltdown. And yet we exist. What new Darwinian mechanism – after double helices and DNA, after polymerase, repair, and proofreading enzymes, after polyploidy and developmental noise reduction – enabled this latest stage of organic complexity to emerge? The answer may lie in the last big unanswered question of evolutionary biology, the ultimate existential absurdity: sex.

5
The ultimate existential absurdity

The information economy

Evolutionary biologists are much teased for their obsession with why sex exists. People like to ask, in an amused way, 'isn't it obvious?' Joking apart, it is far from obvious. The purpose of life is to copy DNA or, to be more exact, information in the form of DNA. Information copying, or information transfer, is a familiar enough activity to us in human culture. We do it all the time. I first wrote this paragraph for a lecture, and then I was vocally broadcasting information, of a sort (it was not information in the form of DNA messages, but in the form of words); the audience was receiving the information, or recreating it – desperately seeking sense in it, in the manner of lecture audiences everywhere. I am now trying to transfer the same information through another medium, the printed page. Information transfer also happens from church pulpits, cinema screens, the stages of theatres. It happens whenever someone speaks and someone else listens, whenever someone writes and someone else reads what was written. It happens when a computer copies a file. Information, in the broad sense I'm using it, comes out of televisions, advertising billboards, holy writ. Human beings have invented an extraordinary range of media for transmitting, or copying, information. But I can tell you one thing about all these media. When humans set themselves the task of copying information, they do just that: they copy it. In biological terms, clonal reproduction (or virgin birth) is the analogy for the way humans transmit information. No one in human culture would try the trick of first making two copies of a message, then breaking each into short bits at random, combining equal amounts from the two to form the version to be transmitted, and throwing the unused half away. You only have to think of sex to see how absurd it is. The 'sexual' method of reading a book would be to buy two copies, rip the pages out, and make a new copy by combining half the pages from one and half from the other, tossing a coin at each page to decide which original to take the page from and which to throw away. To watch a play, you would go twice, pre-programmed to pay attention to the first performance at one random set of times, amounting to half the total length, and to pay attention to the second performance at the complementary other half set of times. You spend the rest of the time switched off. If sex is something of a joke, the joke is not on the evolutionary theorists who have shown it up for what it is.

Sex is not used simply for want of an alternative. Nothing, in an evolutionary sense, forces organisms to reproduce sexually. Indeed, the majority of live reproduction on Earth is probably not sexual. Microbes, such as bacteria, do most of the reproduction on this planet, and they usually do it by doubling their cellular contents and then dividing from one cell to two, without any genetic input from another cell. They clone themselves. Larger, multicellular life forms also reproduce without sex. Some do so by virgin birth. Aphids – which include the greenfly and blackfly that descend like a plague on garden flowers in the summer – can breed by virgin birth; a daughter aphid develops inside its mother and is born without any male input. Plants use another method. They send out a subterranean shoot, which produces new plants here and there; the DNA in the shoot is all produced by simple copying of the genes in the parental plant. I shall refer to all these non-sexual methods of reproduction as 'cloning': all that is required for cloning is that the parental DNA be copied into the offspring. Cloning is simpler than sex, in which the genes of two parents are combined and shuffled, and some of them then disposed of. DNA is routinely copied, without sex, in the cells of our bodies. The alternative to sex is no fantastic illusion, conjured up by mad monks and visionary scientists. Plenty of clonal breeding goes on in the living, breathing world.

The absurdity of sex can be expressed exactly. Imagine a simple population, with three individuals: one clones itself, one is a sexual female, and one is a sexual male. (If the singular numbers sound stupid, multiply each one up to a size that sounds sensible.) Clonal breeders make up a third of the population. The clonal and sexual females are in all respects the same except for the way they breed: they have equally efficient bodies, survive equally well and produce the same number of offspring. They might produce two offspring each, though the exact number is unimportant. The next generation will then contain two clonal individuals, and (if the sexual female produces males and females in equal abundance) one sexual female and one sexual male. Half the population is now clonal. The two clonal and one sexual females again produce two offspring each, and the next generation contains four clonal individuals, one sexual female, and one sexual male: two-thirds of the population are clonal. Before long clonal reproduction will take over. Natural selection will eliminate sex and establish universal cloning. It does so because the clones breed at twice the rate of the sexual females. The ruinous thing about sex is male production. Sexual individuals spend half their effort each generation in producing sons. Males, however, in most species make no energetic contribution to reproduction. Sperm are negligible in size, and contribute only a bit of DNA to the fertilized egg; the existence of males makes no difference to the number of eggs produced. Males use their energies in other ways, such as shouting and posturing – deeply fulfilling actions, to be sure – that do not obviously add to the reproductive output of the female half of the population. If a sexual female were to change over to cloning, she would double the rate of reproduction of her genes. 'I knew sex was expensive,' she might say, 'but I did not know it was *that* expensive.'

Sex is indeed expensive. It halves the rate of reproduction. Its existence is something of a puzzle, because natural selection would not have favoured it unless it had some compensating advantage that at least doubled the quality of the offspring. What could that advantage possibly be? Living creatures in nature are reasonably well adjusted to life; they are, after all, alive. They have the evolutionary option of simply copying themselves and producing offspring that will be (you might think) as well adjusted as they are. An explanation for sex has to show that the offspring of an individual who uses sex will have twice the average quality of a clonal copy of the parent. When biologists think about natural selection, they typically think of things that increase or decrease survival or reproduction by a few per cent or less. Normal evolutionary change probably consists of events that make a small difference. But when we look for an advantage to sex, we have to find something altogether larger. It cannot be any minor fiddle faddle of an advantage. It has to be huge. We therefore might expect it to be obvious.

Martyrs to mutation

Sex is a puzzle that has not yet been solved: no one knows why it exists. Two theories are in favour at present, though both may be wrong. One is that sex helps us in our evolutionary battle against parasites. I have nothing against the parasitic theory – it is a wonderful piece of reasoning and makes sense of many of the crucial facts of life – but it is not my topic here.* The other theory is that sex helps to remove bad genes. The theory is mainly due to an émigré Russian biologist, Alexey Kondrashov, who now works in the United States. Kondrashov's basic idea is well expressed in an analogy by John Maynard Smith. He says: imagine you have two broken-down cars, one immobilized by a faulty ignition system and the other with broken brakes. The two cars are the same make. What should you do? The smart answer is combine the good parts from the two cars in one of the cars. Then you have one car that goes, though at the expense of the second car which is now broken twice over. It does not really matter how many bad parts there are in an already wrecked car; a wreck is a wreck whether it has one, ten or a hundred broken parts. So by sexual motor mechanics – swapping parts between the cars – you can turn two wrecks into one runner plus one wreck, which is a clear improvement.

Sex does the same swap with genes in life. Two individuals might each have one broken gene, but the gene that was broken might differ between the two individuals. If they clone themselves they produce offspring like themselves, but if they have sex then some of the offspring will have both bad genes and other offspring will have neither of the bad genes. The swap helps because some offspring are doomed to die – such is life – simply because parents produce more offspring than on average can survive. It can pay to load as many bad genes as possible into a fraction of the offspring, and sex can be seen as a way of doing so. It increases the number of bad genes removed per death. Whether there is sex or not,

* The theory has been worked on by a number of people, particularly W. D. Hamilton. My near namesake Matt Ridley, with whom I am pleased to be confused, has written a readable book about it called *The Red Queen*. And if two more equally good theories of sex come along, we'll find a Luke and John to write books about them too.

individuals are going to die. But without sex they drop dead with a lower burden of bad genes. Without sex the population, so to speak, throws away its wrecked cars when they have only one bad component. There is a wasted opportunity here. If you are going to die you may as well make a thorough job of it, and take out as many bad genes as possible.

The motor-car analogy expresses the main idea in Kondrashov's theory, but there is more to it. The problem is that the case where sexual motor mechanics helps is only one of many possible kinds of car component (or genetic) swap. We have a total of four kinds of car here. For each car part (ignition and brake systems), there were two kinds, good and bad. If we symbolize a good part by '1' and a bad part by '0' then the car could be 11 if it has two good parts. Or it could have good brakes but bad ignition (symbol 10), bad brakes and good ignition (01), or both parts bad (00). Sex worked in the example because we picked an 01 and a 10 car. They have different bad parts, and a swap between them makes things better on average.

But an 01 car and a 10 car are not the only pair you might pick. Suppose you had started with one car with two good parts (11) and another with two bad parts (00). You make them have sex, swapping parts at random. Sex is now stupid, because you start with a runner and a wreck and end up with two wrecks. With real motor cars you could concentrate sex on the 01 × 10 pairs* with single complementary defects. You can do this because you can tell which parts are defective, and swap only them; you would keep your hands off the 00 and 11 cars. But this option is not available for biological entities. We cannot tell whether we have good or bad genes, and nimbly change between cloning and sex depending on our genetic quality. A good theory of sex should explain how sex works for the average pair, not just a peculiar minority of pairs.

The simple case with two components (or genes) and four kinds of car (or parent) has sixteen possible pairings (Table 4). What is the net effect of sex where all sixteen of these kinds of pairing are going on? The question is easier to answer than may initially appear. Most of the pairs can be ignored, because sex makes no

---

* I shall use the symbol '01 × 10' to stand for a pairing between an 01 and a 10 individual.

Table 4

| | Parent 1 | | | |
Parent 2	11	01	10	00
11	11 × 11	01 × 11	10 × 11	00 × 11
01	11 × 01	01 × 01	10 × 01	00 × 01
10	11 × 10	01 × 10	10 × 10	00 × 10
00	11 × 00	01 × 00	10 × 00	00 × 00

difference in them; the same gene combinations come out of the pair as went in. For a 11 × 11 pairing, a 11 and a 11 set of genes goes in, and a 11 plus a 11 set of genes comes out. The same is true of a 11 × 10 pairing: if you swap the second 1 of the first parent with the 0 of the second, you still have a 11 plus a 10. Sex makes no difference in twelve of the sixteen pairs.

Sex makes a difference only in two kinds of pairs: the 01 × 10 and the 00 × 11 pairs. (The two both appear twice in the table, making up four of the sixteen. The 10 × 01 pairing is genetically equivalent to 01 × 10, and 00 × 11 is genetically equivalent to 11 × 00.) Sex is good in the 01 × 10 pairs and bad in the 00 × 11 pairs. The net effect of sex depends on the relative frequency of these two kinds of pair in the population. If there are more 01 × 10 than 00 × 11 pairs, sex is net advantageous; if there are more 00 × 11 pairs than 01 × 10 pairs, sex is net disadvantageous; if the two pairs are equally frequent, sex is net indifferent. We need to know whether living species correspond to a car pool dominated by total wrecks and perfect runners, where sex makes things worse by spoiling the runners, or by cars with an intermediate number of defects, where sex improves things.

The small fraction of pairings (four of sixteen) in which sex makes a difference is a peculiarity of the two-gene (or two-car component) case. Real organisms have more than two genes, and as you increase the number of genes sex rapidly comes to influence all the pairings. A ten-gene individual would have something like 0111010111 or 1110111100. We have about sixty thousand genes. The number of them that are bad, in an average individual, is not known; but it might be somewhere in the hundred to a thousand range. Then an individual's genes could be represented by a string of a hundred thousand digits made up of a hundred

(or more) os distributed among fifty-nine thousand nine hundred
1s. Different individuals will have different defects, and the number
of ways that a hundred os can be distributed in sixty thousand
digits is so enormous that every individual, and every sexual
offspring, will have a unique set of good and bad genes. Now
there are an astronomical number of possible pairings and the
detail becomes mind-numbing. However, Kondrashov and others
have worked out the theory for the multigene case, and shown
that sex can be advantageous. The reason for the advantage is
conceptually the same as the two-gene case: sex can concentrate
the bad genes in some offspring and the good genes in other
offspring. The advantage of sex actually goes up when we take
more genes into account. With a hundred thousand genes, every
individual is genetically unique and every sexual pairing results in
a change to the offspring genes, rather than just a quarter of the
pairing as in the two-gene case.

Another feature is worth noticing in the realistic, multigene case.
With two genes, or two car components, sex could produce a
perfect individual with no bad genes. In life, probably no individual
is genetically perfect. We all have a few small genetic defects that
make little or no difference most of the time. Suppose, again,
that on average human beings contain 100 small errors scattered
through their DNA. Suppose also that on average two more errors
are added when an individual breeds. With clonal reproduction,
the number of bad genes goes up to 102, and then 104 in the next
generation. There is a mutational meltdown. With sex, the two
parents mix their good and bad genes, and produce some offspring
with more, and others with fewer, than 102 bad genes. All that is
needed is that there are enough offspring with fewer than 100 bad
genes to carry the system forwards with no net deterioration. The
conceptual workings of the theory are again identical in the two-
gene case and the more realistic multigene case. But the theory is
more powerful with the realistic need for enough offspring to be
produced who show no net deterioration from the parental
average, rather than to be error-free. It would be a hopeless task
to produce genuinely error-free offspring by mixing the bad genes
of two parents who contained 100 bad genes each. Fortunately it
is also unnecessary.

The mutational damage escalator

Kondrashov's theory of sex predicts that car pools where sexual motor mechanics prevail will have a certain composition. Sex makes no sense in a car pool in which most of the cars are perfect runners or total wrecks. But it helps where most of the cars have an intermediate number of defects. Which kind of car pools do biological populations correspond to? If you pick two live creatures out of a biological population, do you expect both to have an intermediate number of genetic defects, or do you expect them to have either no defects or many defects? For the present, the answer is unknown. But we do know the conditions that determine the answer.

The key relation is the one between the quality of an organism – how good it is – and how many genetic defects it contains. In crude terms, the more bad genes in an organism, the worse its quality; but the crude relation can have three particular forms (Figure 7). The one that makes sex advantageous is the curve that slopes downwards (labelled a). We are after a population in which it is more likely that an individual contains an intermediate, than an extreme, number of genetic defects. With the curve (a) that slopes down, this is what we get. An individual who contains many defects is of low quality, so low that it will die; natural selection removes these individuals from the population. If an individual has a small number of defects, its quality is as high as an individual with no defects. These individuals with a small number of defects do not die; natural selection allows them to proliferate. Natural selection tunes the population to contain many individuals with few defects, and few individuals with many defects. It is more likely that two individuals picked at random will have an intermediate number of defects than that one individual will have no defects and the other will have many defects. It is the kind of car pool where sex helps. I shall refer to the sloping-down curve (a) in Figure 7 as the 'escalating damage' relation. As the number of defects goes up, each new defect does escalating damage.

The lower curve (c) in the graph has the opposite effect. Here the relation between quality and number of defects goes down steeply to begin with and then flattens off. Natural selection now clears out the individuals with intermediate numbers of defects.

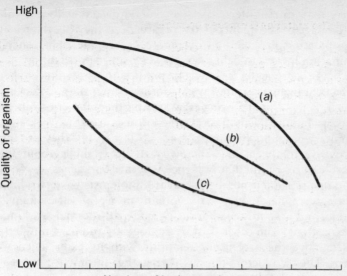

Figure 7. Three possible relations between the number of mutations in a body and the quality of that body. I call the sloping-down curve (labelled *a*) the 'escalating damage' relationship: each successive mutation does incrementally more damage to the quality of the organism. It can be contrasted with the straight line (labelled *b*) in which each successive mutation reduces quality by about the same amount. Alternatively, successive mutations might do decreasing amounts of damage (curve *c*). In a technical model, 'quality' means 'fitness' (the chance that an organism survives and reproduces) and the *y* axis would have log fitness on it. It does not matter which line is above which other line, but it does matter that one (*a*) slopes down, one (*b*) is straight, and one (*c*) flattens off.

They are of low quality, and die. The result is a population in which more individuals have an intermediate than an extreme number of defects. Sex is now disastrous as a way of clearing bad genes: each death removes fewer bad genes than if the population cloned itself. Finally, with the straight line (*b*), natural selection makes the frequencies of the 00 × 11 and the 01 × 10 pairs equal. Sex hinders as often as it helps. With either of the lower two lines, natural selection does not favour sex. Kondrashov's theory of sex predicts that, in real life, successive genetic defects will do escalating damage to the quality of the organism. If they do, the result is a population in which an average pair has the right genetic combination for sex to help them produce superior offspring.

We have theoretically replaced one prediction, about the number of defects in living creatures, with another, about the relationship between the quality of a living creature and the number of genetic defects it contains. This second prediction is testable, and some preliminary attempts have been made to test it; but they are too inconclusive to dwell on. What we can do is think about the general features of life that influence whether bad genes do escalating damage as their numbers increase. Biologists have two lines of thought. One imagines the organism as a self-contained machine, and asks how bad genes damage the internal workings of the machine. The other is more ecological, and asks how bad genes damage the organism's relations with its external environment. The two factors may work together in practice but it is easiest to consider them separately.

In Chapter 4, we looked at various developmental troubleshooting mechanisms. The gene $p53$, for example, prevents cells with mutant DNA from damaging the body: it makes the cell commit suicide. The body can tolerate some mutations because $p53$ stops the mutations from doing much harm. But what if the $p53$ gene itself suffers a mutation? This is much more dangerous because other mutations can now do their worst. In the case of a mutated $p53$ gene, the body will probably die of cancer. Successive mutations are likely to do escalating damage, because the first few are unlikely by chance to hit a troubleshooting gene. But as more and more mutations arise, a troubleshooting gene will eventually be hit. The quality of the body will then decline dramatically.

We can extend the same idea to another, probably more numerous, class of genes: genes coding for back-up mechanisms. When one bodily mechanism is knocked out by mutation, the body often has some other mechanism that can do the job almost equally well; the mutation therefore does little harm. For example, most of us have two functioning eyes. Imagine a mutation that removed one of them, the left eye. Eyesight quality would be lowered; the mutant person's perception of distance would be impaired and part of the field of vision would be lost. I can estimate the amount lost by closing my left eye: about a quarter of the field of vision disappears if I do not move my head. A 180° field of vision is reduced to about 135°. I can compensate by rotating my head, but this takes time and thought and I lose vision on the other side as I do so. If a tiger approached in the left-hand 40–50° blind zone, I should have a higher chance of becoming tiger food than if I had both eyes. My chance of survival would be slightly reduced. The complementary story could be told for the complementary mutation which removed my right eye. But if I lack both eyes, I deteriorate by much more than the compound effect of losing each individual eye. Now I am blind and (in a state of nature) probably dead. The relation between the number of defects and body quality slopes down. The reason is that one defective organ – such as a lost eye – can be backed up by another organ. Back-up systems are common in life. We have two ears, kidneys, lungs, arms and legs, and one of them can back up for the other just as with eyes. To lose one kidney is a misfortune, but to lose two is worse than carelessness: it is death. The chance of survival of a body will slope down as the number of eyes, or of kidneys, decreases one by one.

However, duplicated organs such as eyes and kidneys do not completely make the case we need. Successive losses in these organs do cause escalating damage to the body, but they are probably not coded for by separate genes. It probably is not true that one mutation can knock out the right eye and a second mutation the left eye. I want now to turn to some other evidence, which definitely concerns genes, though the case that they can back up for one another is less certain than for eyes. The evidence comes from duplicate genes. The DNA of an organism may contain several genes that are very similar to one another. The genes probably originated by the kinds of duplications we looked at in Chapter 2, and they are called duplicate genes. If there is one thing that has been shown by the genome projects of recent years it is that the DNA of all life forms abounds with duplicate genes. In 1996, for example, a team of 600 scientists completed the sequence of the yeast genome. About 1000 genes were known in yeast before the sequencing project, but the full sequence contained 6000 genes. The function of most of the new genes is unknown, but many of them appear to be duplicates. In 1997, another large team finished sequencing the bacterium *E. coli*. It has 4000 genes, and 40 per cent of them are of unknown function; about a third of the genes look like duplicates. In 1998 the 'C. elegans sequencing consortium' published a DNA sequence for the worm. They found 19,000 genes, more than had been estimated from earlier work. The multicellular worm has about three times as many genes as the single-celled yeast, and much of the increase seems to be because the worm uses larger families of related (and duplicate) genes than does yeast. Some of the extra genes in the worm must also control development. We are in the early days of DNA sequencing research, but the typical pattern so far has been for the DNA sequence of a species to reveal a host of previously unidentified genes, of which many are duplicates of other genes.

The blood oxygen-carrying molecule haemoglobin illustrates how duplicate genes may back up for one another, and how defects in duplicated genes may do escalating damage. Haemoglobin is a tetrameric molecule, made up of four globin molecules: the haemoglobin of an adult human being is made up of two α-globin chains and two β-globin chains. The four globin molecules are coded for by four separate genes, and there are other globin genes too. If a body contains a mutation in one of the four genes, it can make a substitute haemoglobin using the products of other globin genes. People who lack globin genes suffer from the disease called thalassaemia, and there is a whole array of thalassaemias depending on which globin genes are lost and which other globin genes are used instead. People who lack β-globin are said to have β-thalassaemia, and in one kind of β-thalassaemia you continue to use fetal haemoglobin after birth. Fetal haemoglobin is a tetramer of two α-globins and two γ-globins. The γ-globin gene is yet another member of the globin gene family, and is normally used only in the womb. Fetal haemoglobin seems to work almost as well as adult haemoglobin after birth. Adults who use fetal haemoglobin are described as 'symptomless': they are indistinguishable from normal adults. There are some small differences in the properties of their red blood cells and they may suffer under some circumstances such as mountaineering; but γ-globin seems to be a good substitute for β-globin.

There are also various kinds of α-thalassaemias, in which the body lacks α-globin genes. A normal individual has a total of four α-globin genes, and there are four degrees of α-thalassaemia, depending on whether you lack one, two, three or four α-globin genes. If you lack one, the condition is symptomless; lack two, and you have some minor abnormality in your red blood cells but are otherwise fine; lack three and you are severely anaemic; lack four and you are dead at birth or soon after. The missing α-globins can be made up for in several ways. Among the best is to produce haemoglobin H, a tetramer of four β-globin chains: it does a passable job, but is much better at binding oxygen than releasing it and has other shortcomings. The various globin genes, therefore, can to some extent stand in for one another. The same story can be told for myoglobin, which is yet another member of the globin gene family and specializes in carrying oxygen in muscles. Mice that have their myoglobin gene experimentally knocked out are physiologically almost indistinguishable from normal mice – presumably because some other globin gene stood in for it.

The globin genes work as specialists, each adapted to carrying oxygen at a particular developmental stage, or in a particular region of the body, or combining with other globins to make a superior compound molecule. But they all have some oxygen-carrying ability in all circumstances and if one of the specialists is lost, another gene can stand in. Successive mutations in the globin genes probably do escalating damage. If you lose one globin gene you can still function normally (except in conditions of oxygen stress); if you lose two you can probably work adequately most of the time; if you lose three it may start to hurt (depending on the three that go); if you lose all four you are dead. The globin genes, therefore, may resemble eyes and other back-uppable organs. The body can cope with a few defects, but not with many. The globin genes are particularly important because they may illustrate a general feature of duplicate gene systems, and our bodies undoubtedly contain many families of duplicate genes. They point to a general theory of escalating damage, a theory that follows from the way our bodies are built, and work, as machines.

The second theory almost ignores the internal workings of the bodily machinery. It concentrates instead on how individuals compete for resources. The competition for territories makes a concrete example. Many birds compete for territories, and only the birds who successfully occupy a territory may manage to breed. In a simple case there might be ten territories and twenty birds. If all ten territories were equally good, then the ten strongest birds would all breed, with equal success, and the ten weakest birds would not breed at all. We could rank the twenty birds according to their strength, from bird number 1, who is the strongest, to bird number 20, the weakest (I am assuming that territory ownership depends on strength). If we look at the relation between rank and reproductive success, we find that reproduction is high for birds 1 to 10 and low for birds 11 to 20. If the territories differ in quality, bird 1 would take the best territory and bird 10 would be left with the worst territory. The relation between rank and numbers of chicks produced would be less extreme than when all territories were equally good, but still there would be a decline between the territory owners and non-owners.

Now for the crucial step in the argument. We shall assume that a bird's strength depends on the quality of its genes. The assumption is realistic because mutations could reduce a bird's strength in many ways: a bird could be weakened by a mutation in a gene that coded for muscle, or bone, or ability to carry oxygen to the muscle, or ability to digest sugar, or to find food, or many other things. All we need is that there is some tendency for bad genes to reduce the competitive power of a bird. Bird number 1 will have the fewest bad genes, bird 20 will have the most bad genes. Successive mutations will then do escalating damage to these birds. Take a bird who has few bad genes, and is in the top ten and doing well in life. If you introduce successive mutations into this bird, at some point its strength will decrease below that of bird 11; it will be relegated to the ranks of the non-owners, and its output of chicks will decrease to zero. The escalating damage has nothing to do with back-ups or other interactions between genes inside the body; it is caused by the resources and the way they are competed for.

The exact way that gene quality influences strength in an absolute sense (such as might be measured by the avian equivalent of weight-lifting) does not matter. It may be that successive mutations decrease strength by constant amounts, or by diminishing or escalating amounts. These are details we can ignore: what matters is the competition between birds. If the number of bad genes in a body influences competitive power, and if the competition is for resources that divide the competitors into haves and have-nots, then the relation between mutations and reproductive success should show escalating damage. Sex will be advantageous.

I have described two reasons why bad genes may work in the way Kondrashov's theory requires. Multiplying defects may eventually wreck the internal workings of the body, or they may eventually prevent a body from being able to obtain limited resources, or both. Ultimately it will take facts to decide whether bad genes really do more damage as their numbers increase. These facts will be important, because they are crucial for our understanding of complex life. In Chapter 3 I argued that a harmful mutation rate of roughly 1 per offspring is the upper limit for life. The evidence in Chapter 4 suggested that mutation rates may have reached this limit in worms and flies, implying a ceiling on live complexity somewhat below the level of worms. And yet, paradoxically, life forms of greater complexity have evolved, including ourselves with our mutation rates unambiguously higher than 1. Sex is one possible solution. Sex can smash the ceiling on the error rate, and on the complexity of life. The upper limit of 1 appeared to be logically solid in Chapter 3 because we did not think about recombining the errors after they had arisen. The information transfer in the Chinese whispers game is clonal. One child receives a message from a child earlier in the line, and transfers it on to one child later in the line. No child tries anything 'sexual' such as mixing two messages and transferring the scrambled version on. With clonal reproduction, natural selection is overwhelmed when every recipient has an erroneous message. The trick of sex is to combine errors from different individuals and create some error-free individuals. The mutational sins of the parents are concentrated in a fraction of the offspring. Multiple mutations can then be removed by one death, and an error rate of more than 1 becomes tolerable. If an average death eliminates ten mutations, the upper limit on the error rate goes up to about 10. The mutation rates that I described in Chapter 4 as 'paradoxical' need not be paradoxical after all: life forms can have a mutation rate of more than 1 provided that they also breed using sex. However, sex only helps to purge mutations if successive mutations do escalating damage to the body. The arguments I made about back-up mechanisms in eyes and duplicate genes, and about competition for territories, were both just that – arguments. They could turn out to be wrong. Maybe successive mutations do not do escalating damage. We should then be back with the case in which swapping genes between individuals makes things worse as often as it makes them better. Sex would not act to concentrate errors, and the logic of Chinese whispers would be unaltered. For now, Kondrashov's theory is at the testing phase. It has suspended, but not finally solved, the paradox of life forms with high error rates.

The moonlight sonata

Animals, particularly male animals, do some weird things before they breed. Songbirds sing, frogs croak, fruit flies dance and throw up. Peacocks grow a fan of deeply coloured feathers with images of eyes painted on them; male butterflies have evertable and perhaps scented plumes which they spring out during courtship; stalk-eyed flies have their eyes at the end of stalks that are longer than their bodies. The standardwing bird of paradise lives in the northern Moluccas, Indonesia, and the 'males, on the approach of a female, go into a frenzy, leap vertically up to 11 metres into the air, and descend parachute-like with their four long white wing standards flared.' It is all most curious. The explanation is probably, as Darwin first realized, mate choice: females choose to mate with the males who perform the displays, and perform them most energetically or extensively. But Darwin's solution replaces one puzzle by another: why, we now ask, did females evolve to choose their mates by these criteria? Biologists have offered many answers to this question but here I wish to concentrate on one possibility – that the displays act as indicators of gene quality. (I shall express the arguments in terms of female choice, among male suitors. Most mate choice, in most life, has this form. In human beings, however, there is pair-bonding and some male choice among females as well. The biological problems are the same either way. Also, the word 'choice' refers to behaviour and implies nothing about consciousness. If two kinds of male are available and females tend to mate with one kind rather than the other, the females are said to have 'chosen' the males they mate with.)

Reproductive partners with superior genes are evolutionarily desirable, because they produce superior offspring. The hard part is to find out the genetic quality of a potential partner. It does not come written on him – the number of his bad genes is not inscribed on his forehead. Nor can his DNA be read directly, in the manner of a proofreading enzyme, and the errors counted there. Genetic quality has to be assessed abstractly and indirectly, from the individual's appearance and behaviour.

The problem is like taking a car out on a test drive. You know little – at least in the biological analogy – about how cars work, and you lack the skills and the facilities of a motor mechanic; you cannot assess the car's components directly. Your aim is to avoid buying a dud, or 'lemon'. The first thing to establish is that it goes at all. You should try it under demanding conditions, including flat-out acceleration in a safe place and intimate manoeuvres in traffic jams. The car should be able to give you a smooth and safe, or at other times exciting, ride. You do not want it to leak dirty fluids or to backfire, but it should make a good loud noise when you sound the horn – when you are angry you will want it to make people jump. You should try all the buttons, switches and levers on the control panels, and check that they produce the right response. You might also kick the tyres. Being courted is a 'pro-active' experience.

Your mechanical ignorance is one problem; another is that you are going to be accompanied by a smooth-talking salesperson, who will assist you in your decision. He or she knows more about the car than you do and wants to sell it, independently of its quality. The salesperson is rather like a courting male, who may privately be aware of some of his defects but whose interest is closure, not candour. If you are innocent about cars you go through a routine I know all about. 'The revs sound a bit odd,' you say, which is a cue for something like, 'Oh no, Dr Ridley, the new Fords are equipped with ergonological gearing, and they sound like that ... it is an advanced feature [etc.].' If a courting male analogously misfires – back comes the credit card, on a plate, neatly cut into two pieces – he will be ready with plausible excuses. The salesperson can also act to conceal defects in the car. If the car is one that gasps and wheezes at high speeds, he can point out that there are roadworks on the motorway today and it would probably be better to avoid it. A courting male may also know that his machine code is corrupted somewhere in the tiger-defence instructions. He may then invite you to an alternative treat, on the day he suspects that you intend to test his performance in the tiger cage. A male who has managed to survive may have done so by having good genes, or by having poor genes and looking after himself. Ideally you need to take him out and rev up every component, one by one and in combination, to test for defects; but this is an exhausting job in a 60,000-gene life form.

Courtship may have evolved to allow a fast assessment of genetic quality, in a way that is not open to deception. Courtship consists of dance steps, ritualized displays, and animal noises. Evolution may have ensured that a male with accurately copied genes can grow the tail, or dance the dance, or sing the song, whereas a male with corrupt genes cannot, or cannot so well. An example comes from the creatures of fairy-tale romance – frogs. Courtship in most frogs and toads is a musical affair, in which the male attracts visiting females by means of calls, or croaks. The female grey tree frogs of Missouri in the USA choose the males with the longest-lasting calls – both in terms of the number of call 'pulses' and the length of each pulse. Alison Welch, Raymond Semlitsch and Carl Gerhardt at the University of Missouri did an ingenious experiment to test whether the preferred males were genetically superior. They crossed each female with two males, one with a long call and the other with a short call. Some of her frogspawn was fertilized by sperm from a long-call male and the rest by sperm from a short-call male. They did this for twenty females, using fifteen male pairs. It is a neat trick to use frogspawn from one female with more than one male, because it controls for any peculiarities of the individual females. Welch, Semlitsch and Gerhardt then measured various attributes of the resulting tadpoles, such as how fast they grew and how well they survived. They measured the attributes under two conditions, one with plenty of food and the other with a limited food supply. The result was that (for the various attributes and measuring conditions) either the tadpoles fathered by the long-call males were superior to those fathered by the short-call males, or there was no difference. The net result of all the measurements was a clear superiority of the tadpoles fathered by males with long calls. The experiment is so well designed that the superiority almost has to be due to the quality of the male genes. Non-genetic influences, 'maternal effects' and the genetic quality of females are all controlled for; the use of a range of measuring conditions suggests that the result should apply in natural frogs as well. The male frog's call, therefore, works as an indicator of the underlying quality that really matters: the quality of his genes. It is as if you could assess the quality of a car simply by listening to its engine.

Grey tree frogs are not the only species in which it has been shown that genetic quality assessment goes on during courtship, but the evidence for this species is exceptionally strong. There is also nice work, by Jerry Wilkinson and various colleagues, on the bizarre stalk-eyed flies. (Wilkinson is based at the University of Maryland in the USA, but the flies themselves live in Malaysia.) Female stalk-eyed flies choose the males with the longest eye-stalks, whose eyes are furthest apart. It turns out that these males are less likely than average to contain a certain kind of rogue gene – a 'lawbreaking' gene in the sense we shall meet in Chapter 7 – that reduces the quality of its bearers. There is even some possible evidence from human beings, as we shall see in Chapter 9. I find it an amazing idea, that courtship and sexual signalling in all its variety could have evolved to reveal the error content of male DNA. Peacocks with corrupt codes may print out small, dull and blurry feathers. Birds of paradise with corrupt codes may crash instead of parachuting, or manage only a pathetic little jump. Our own sense of beauty may partly have its origin in genetic inspection. Our art and music markets would then exist not only for aesthetic reasons, status signalling and mood manipulation, but also because beauty evolutionarily guarantees the copying accuracy of life's internal DNA codes. Accurately copied DNA is the great aphrodisiac.

The advantage of mate choice is related to the advantage of sex itself, in Kondrashov's theory. Mate choice adds an additional advantage, if the theory is correct; but even if Kondrashov is incorrect and sex exists for some other reason it still pays to pair up with someone who has superior DNA. The theories of sex and of mate choice face a common problem and solve it in different ways. The problem is to ensure that, with all the good and bad genes hanging out in the population, the offspring of a sexual pair tend to inherit the good rather than the bad genes. The solution is easier in the theory of mate choice. It assumes that individuals can detect gene quality. The offspring then inherit good genes, because the males who have good genes do a disproportionate amount of the fathering. In Kondrashov's theory, the individuals cannot detect the quality of another individual's genes. Sex has to work even though males and females pair up independently of their genetic composition. That is the problem ingeniously solved by the 'escalating damage' relation: it ensures that the net effect of sex is to concentrate the bad genes in a minority of unfortunate offspring. Either individuals cannot distinguish good and bad genes in members of the opposite sex, in which case something special has had to act to ensure that sex works even with random pairing, or individuals have to be able to assess genetic quality.

The evolution of a genetically efficient mating market was a new stage in the struggle of complex life against copying error. It must have followed the evolution of sex, because mate choice is meaningless in a clonal life form. In evolutionary terms, mate choice helps by focusing natural selection to act more powerfully against genetic error. Natural selection automatically removes harmful mutations, but some do survive; now they have to pass a second test before they can make it into the next generation: testing by female sensory systems. The motor-car analogy can express the point. The cars in an area can be taken off the road if they break down or have an accident, or if they fail an official certification. The removal of cars by accidents and breakdowns is like death; it is analogous to simple natural selection. But accidents do not remove every ropy car, any more than premature death removes every male with ropy genes. The evolution of mate choice was analogous to a law demanding that every car be regularly inspected by experts. Now the cars with detectable defects can be taken off the road even if they have not yet broken down. When females evolved the ability to detect gene quality, the purge of bad genes grew more efficient, and a new stage of female quality control was added before each new generation of machines was run off the genetic assembly lines.

Life has evolved a series of mechanisms to recognize bad genes. We have molecular mechanisms that recognize errors in the DNA, and correct the errors. We have internal troubleshooting mechanisms that recognize at least the effects of bad genes, and fix the effects but not the genes. Natural selection has also, it seems, incorporated an indirect, intuitive bad-gene detector into the acoustic systems of grey tree frogs. In the past, female frogs probably responded to all sorts of stimuli; but only those who approached the males that called long and often became the ancestors of modern frogs. The love song of the grey tree frog now displays the singer's genetic worth to every listener. When the croaks pulse fast and long, it shows that the male genes are pure, and the frog princess prepares, in the Missouri moonlight, for the ceremony of the frogspawn.

The original recreational sex

Bacteria are mainly clonal, but they also use mock-sexual processes from time to time. Bacterial sex is interestingly different from the kind that we and all complex life forms use. We swap genes: a male and a female each contribute half their genes to an offspring. Bacteria do not swap genes: a bacterial sex act has a gene donor and a gene recipient. Two bacteria come together, build a connecting bridge, and one bacterium sends genes over the bridge into the other; the two bacteria then separate. The bacterium that received the genes may then add the new genes to its existing supply, or replace some of its own genes with the donor's genes. The donor bacterium is unaffected. It just makes an extra copy of some of its genes and injects them into the recipient; nothing happens to its own genes. Bacterial sex is not connected with reproduction. There were two bacteria before the sex act, and two after it. Both may later go on to reproduce, but that is an independent process from sex. When a bacterial cell reproduces, it just copies its genes and divides in two. It divides in the same way whether or not it has received another bacterium's genes earlier in its cellular life. Bacterial sex is rather as if we human beings could take in genes from another individual, in some or all the cells of our bodies, during our lifetimes. I might benefit from someone else's genes in my arm, and that other person would then make extra copies of his or her genes and somehow transfer them to my arm cells.

It is not known how bacterial sex fits in with, or challenges, the story of this chapter. The circumstances, and advantages, of bacterial sex remain too mysterious. The first step is for a bacterial cell to undergo special changes which make it receptive to the genes of other bacteria. It is uncertain why the bacterial cell undergoes these changes but one theory is that it does so when it is in difficulty. It may be sick, or struggling, for instance because it is being blasted with antibiotics. Experiments show that bacteria are likely to become receptive after their DNA has been damaged. It is as if the bacterium becomes receptive when it 'knows' that it has worse than average genes. The bacterium can then gain by taking in some genes from any old other bacterium, with average-quality genes. There is something rather like 'mate choice' going on. Sex can help if it results in the transfer of genes from average individuals to below-average individuals. A bacterium probably cannot discriminate between potential gene donors, by some means analogous to music, as frogs do (but who knows – maybe they can). However, sex still helps the recipient provided that it is genetically below average. It could detect that it is below average by some crude evidence, such as that it is struggling to stay alive, or more precise evidence, such as damage in its DNA.

In the bacterial car pool, sexual motor mechanics may be acting to replace inferior by average components. Bacteria would then not need to show the 'escalating damage' relation in order for sex to be advantageous. The escalating damage relation was needed when the components were randomly swapped between cars. But bacteria may not be swapping components at random; they may be selectively replacing bad components with good. Their pairing process then itself acts to purge bad genes. The difference from our kind of sex is that we do not have separate donors and recipients. When a sperm fuses with an egg, the male and female contribute equally to the offspring; our kind of sex is genetically egalitarian. It is not the case that, if one of the partners is genetically inferior to the other, the superior partner contributes more than 50 per cent of the genes to the offspring and the inferior partner less than 50 per cent. Our sex act is a commitment in both partners to use 50 per cent of the genes of the other individual in reproduction, for better or worse. In bacteria, the donor cell would gain no advantage if it had to swap its genes with the inferior, recipient bacterium. The donor gains by copying its genes into the recipient, and evolutionarily does not care how corrupt the genes are in the other bacterium before the sex act. But it would care a whole lot if it had to take some of the recipient's genes in return.

In this theory, both the donor and the recipient bacterium gain from sex. The donor has copied its genes into another individual. The recipient gains because it is inferior and the imported genes save its life. The recipient so to speak re-creates itself, and bacterial sex, which is connected with personal development but unconnected with reproduction, may be seen as the original recreational sex. The theory makes sense and has some evidence in its favour; but it is far from proved. It would be premature to draw any firm conclusions about the influence of bacterial sex on the purge of bad genes.

The history of life and the history of sex

We saw in Chapter 4 that bacteria have harmful mutation rates well below 1 and, as gene numbers have expanded in the increasingly complex forms of life, so the harmful mutation rate has probably climbed above 1. The more complex the life form, the more ferocious the headwind of mutations that it has to persist against. Moreover, simpler forms of life tend to have short generations and high fecundity, giving natural selection more scope to purge mutant organisms than in complex life. Complex, multicellular life has more need of sex, and we can predict that it will make more use of it. The prediction is crude, but we can test it against the general distribution of the sex habit in the tree of life. I shall start with the complexity range from bacteria to us, and then come back to viruses.

The facts fit the prediction rather well. Bacteria do use sex, or mechanisms rather like sex, as we have just seen; but bacteria are mainly clonal. When a bacterium divides it usually passes on copies of the same DNA as it inherited. They do suffer from bad genes, and may sometimes have so many bad genes that sex becomes advantageous, but most of the time it is better for them to clone themselves. Likewise, single-celled eukaryotes do use sex, but probably mainly breed clonally. The picture for multicellular animals can be seen in Graham Bell's book *The Masterpiece of Nature*. Some kind of cloning is found, along with sex, in all the simpler animals – sponges, jellyfish, corals, all kinds of worms, rotifers, polyzoans and waterfleas. Cloning is also scattered among the huge ranks of insects, though most insects use sex. Only when we turn to vertebrates does clonal reproduction become truly rare. A few fish, amphibians and lizards do breed clonally, or in ways very like it; but no bird or mammal naturally reproduces without sex. In the animal branch of the tree of life, it is indeed the most complex forms that rely the most on sex. Plants probably show the same trend, though it is difficult to discern in the encyclopaedic variety of their sexual practices.

And what about viruses, which (as we also saw in Chapter 4) include forms that have mutation rates near the limit of the clonally possible? The RNA viruses range in length from less than 10,000 to more than 30,000 letters. It is the longer viruses that are nearer the limit, and the longer RNA viruses may therefore need to shuffle their errors sexually in order to continue to exist. There is some encouraging evidence in support. Longer RNA viruses do indeed use (or appear to use) sex more than shorter RNA viruses. Viral sex is an obscure topic, and surrounded by a haze of scientific uncertainty. I am not sure it will turn out that RNA viruses need sex to perpetuate themselves despite their accident-prone self-replication. But if they do, we can imagine a particularly charming cure for HIV infection and AIDS. We just need to put a stop to sex in the creatures – metaphorically throw a bucket of cold water over them – and they will be forced into clonal reproduction and an inevitable mutational meltdown. Deprived of sex, the plague of a once sexually transmitted disease will become a de-sexed, un-transmitted, ex-disease.

The biologist J. B. S. Haldane once remarked that 'evolution could roll on fairly efficiently without sex but that such a world would be a dull one to live in.' The attribution may be apocryphal (the shade of Haldane is a strange attractor of questionable stories), but it is certainly true in spirit. Bacteria can probably thrive with little or no sex. The history of life has been dominated by these creatures and it would hardly be surprising if they were the only life forms to exist on Earth. But with bacterial error rates, and clonal reproduction, life is limited to single cells. It could hardly stretch to a multicellular eukaryote, let alone the all-singing, all-dancing fabulous extremes of live complexity. The complexity of life has expanded beyond bacteria, and the error rate has increased to a higher level than was tolerable under the old clonal regime. As these forms (including ourselves) evolved, sex was no longer just occasionally useful: it was obligatory.

The evolution of sex led to the evolution of genetic dis-crimination in the mating market. However, although we know the approximate distribution of sexual, as opposed to clonal, reproduction in the tree of life, we do not know much about mate choice. We saw how bacteria may go in for a kind of mate choice, if genetically inferior bacteria are the gene receivers in bacterial sex; but that was mainly a conjecture, and in any case becomes impossible after the evolution of the kind of egalitarian sex that we and all but the simplest life forms use. In us, males and females contribute equal sets of genes to the offspring. The only way that inferior genes can be discriminated against is for their bearers to lose out in the mating market, before any sex act takes place.

The discrimination is executed during courtship. In complex life, courtship procedures may have evolved to expose corrupt DNA more ruthlessly than in the simpler courtships of simple life forms. The courtship of birds certainly goes on longer than the courtship of bees. The evolution of complexity may have advanced hand in hand with the evolution of genetically discriminating courtship. Live complexity may have evolved to become the destroyer, as well as the creator, of mutational error.

Imagine you are God and your chosen people are sinning too much. Sins are rather like mutational errors if you are trying to preserve a certain moral, godly order in a society. What should you do? One possibility is what Christians would call the Old Testament method. You lose your temper and kill the sinners. If the sin rate zooms up to about one per person – everyone is a sinner – then you have to kill the lot: drown them with a deluge, or blast them, like the citizens of Sodom and Gomorrah, with fire and brimstone. 'Now,' you thunder at them, 'Now you know the meaning of a mutational meltdown.' As life has evolved through bacteria, and simple multicellular forms, natural selection has been able to maintain order by some combination of high fecundity, improved accuracy of DNA copying, and multiple gene sets. But at some stage all these were no longer enough. Gene numbers grew so large and error rates so high that a limit on the complexity of life was reached. I do not know what that stage was – maybe flies, maybe fish, maybe you and me. By the time we reach humans, simply copying the genes and killing the bad copies is unlikely to be enough. There are just too many mistakes. Evolution then moved on to the New Testament method. You take as many sins as possible, combine them in one or a small number of scapegoats, and crucify the creature, taking out multiple sins in one individual. Life does this with sex – or may do it with sex, if the necessary conditions are satisfied. Perhaps it is another of those theological absurdities that when the Christian God introduced his new policy for sin – redeeming it with love, not eliminating it with death – the multiple sin-ridden scapegoat was, as we are assured by the Septuagint, created clonally, by virgin birth, in an immaculate conception.

6
Darwinian mergers and acquisitions

The aftermath of a megamerger

The evolution of the eukaryotic cell looks, with the benefit of hindsight, like the big breakthrough in the history of complex life. All life on Earth (except viruses) is built of either prokaryotic or eukaryotic cells. Prokaryotic cells are smaller, have simpler internal machinery and contain fewer genes. Bacteria are single-celled prokaryotic creatures and use about 2500 genes. Eukaryotic cells are in all respects more complex, and even a simple, single-celled eukaryote such as bakers' yeast uses 6000 genes. Practically all multicelled life on Earth is built from eukaryotic cells. Every human cell is eukaryotic. Moreover, the origin of the eukaryotic cell was either the rate-limiting step in the rise of complex life on Earth, or one of the two main rate-limiting steps, as we saw in Chapter 1. Life on Earth originated rapidly but then remained simple for almost 1500–2000 million years until the evolution of the eukaryotic cell. Multicellular life forms, with complex egg-to-adult development, may then have evolved either rapidly or after a delay, depending on which dates turn out to be right. The relative times of the origin of life, the origin of the eukaryotic cell and the origin of multicellular life suggest that the eukaryotic cell was an improbable step.

The eukaryotic cell evolved in a merger between two kinds of bacterial cell. The merger itself was not, in an exact sense, the origin of the eukaryotic cell. The defining difference between a prokaryotic and a eukaryotic cell is the absence or presence of a nucleus. The DNA of a eukaryotic cell is contained in a special compartment, called the nucleus, surrounded by a membrane. The DNA of a prokaryotic cell lies uncompartmentalized in the cell. If you claim to be talking about the origin of the eukaryotic cell, your topic probably ought to be the nucleus. But I am not concerned here with why eukaryotic DNA is wrapped in a nucleus; I am purely concerned with the extra genes that eukaryotic cells contain relative to bacteria. The merger between two or more prokaryotic cells did cause an increase in gene numbers, even if it did not instantly create the cell nucleus.

The initial merger probably happened when a large bacterial cell engulfed another, smaller bacterium. The cell-within-a-cell combination may have been more efficient, or powerful, than either cell alone, perhaps because the two partners had complementary skills. In modern eukaryotic life forms, such as ourselves, the descendants of the engulfed cell still exist in the form of our mitochondria. Mitochondria are rod-shaped structures that are present in almost all our cells. They are the cellular furnaces that burn fuels in oxygen and generate energy. Mitochondria are a major competitive advantage for eukaryotic cells, because they are twenty times as efficient at producing energy as are bacterial cells. The original merger partners may have worked well together because one could produce fuel that the other could burn, but the exact operations of the ancestral cell are not important here. What matters is that the cell probably had some synergetic advantage, and that it contained extra genes because it combined the genes of two cells.

Modern mitochondria are descended from an ancestral bacterium that was once free-living. We do not know how many genes it had, but it might have been a typical bacterial figure of about 2500 genes. After it merged into the eukaryotic cell, the bacterium-mitochondrion shed most of its genes, transferring some of them to the nuclear DNA; but it retained a few genes, which are still present in modern mitochondria. Human mitochondria, for example, like those of all animals, contain thirteen genes that code for proteins and twenty-four genes that code for RNA. A human cell therefore contains two separate sets of replicating DNA. One is in the nucleus and contains about sixty thousand genes; the other is in every mitochondrion, and contains thirty-seven genes. The mitochondrial genes are vestiges of a 2000-million-year-old merger event. Some plant cells contain additional structures called chloroplasts, which are the sites of photosynthesis. Chloroplasts also contain genes, and are also descended from an earlier merger event. In this chapter we shall be thinking about how natural selection acts during mergers of the kind that created the eukaryotic cell.

In business affairs, it is common for two companies to cooperate temporarily on some contract. A business might hire a caterer; a store chain might subcontract its transport; an architect might work with a builder. They need ample discussion first, and legal safeguards, and even then disputes often break out. But cooperation is a common kind of interaction. It is 'easy' or 'probable', given the way the world is. Complete takeovers are less common, but they happen often enough. One management rules the whole, the other management is let go. Problems often arise in stitching together the two enterprises, but they are dealt with in some form or other. In a takeover, the two managerial power systems are never both retained. At best, dual control would create problems of communication, with the possibility of errors and delays. More realistically, the two managements would compete, and more or less subtly undermine each other and direct resources into their own, rather than the other management's, domain. A business is less efficient if it has competing factions within, and it needs to be structured in such a way that internal squabbling is minimized. This is why one management has to go, or be completely subordinated, in a business takeover. Businesses are in such a hurry to remove the managers that they may even be put out before the companies are formally acquired. They may then make off with the boardroom pictures and the company car, but have little time to do strategic damage, such as establishing contacts for the future business they intend to start up.

What is seriously difficult to do, and happens only rarely, is a merger of equals, in which two businesses with separate histories are put together with both managements in place. Even when this does happen, it typically leads in a few months to a thinly concealed row, after which one management resigns 'to pursue other business interests' (in the British euphemism) or 'spend more time with their families' (in the American euphemism). In the merger that led to the eukaryotic cell, there was a successful merger of equals even though two managements were put together, each with their own compensation committee answerable to themselves. Both managements – both DNA sets – had to be retained because they needed each other. Each managerial team had skills that the other team could not provide. It is as if a German business and an American business merged and neither management was confident it possessed the linguistic and business-culture skills of the other country. Neither management wanted to sack the other management as a whole. But both would find some competitive opportunities. The American managers might favour higher investment in the plant under their control, even if it reduced the profitability of the whole business, provided that it raised their own pay. Or, over time, the Germans might find that they really could do what they formerly thought that only the Americans could do, and they would favour expanding the German managerial domain at the expense of the American. It will be a wonder if the business survives with its original dual command structure. But when strange, innovative deals are done, they can have strange results. Somehow, in the evolution of the eukaryotic cell, the forces of managerial competition were tamed and a cell type evolved that was more complex – contained more genetic information – than any cell on Earth before.

The ancestral eukaryotic cell was a set-up for a Darwinian post-merger bloodbath. The merged cell would initially have contained two gene-managerial systems, one from each of the two former bacterial cells (Figure 8). When the cell came to reproduce, the two sets of genes should have been copied into the daughter cells. If the two gene sets were about equal in size, and copied in fair (I mean equal) amounts, they would still be in a 50:50 ratio when the cell reproduced. But natural selection would strongly favour either gene set if it could copy itself in unfair amounts within the cell. One of them might copy itself more rapidly, proliferate in the cell, and be more likely to be passed on during the next round of reproduction. Selection then acts on the other gene set to counter-proliferate, or make the proliferating DNA slow down. The force of selection here is strange. In the famous examples of natural selection, one kind of organism is favoured over another. Dark moths are favoured over light moths, in polluted areas where dark coloration provides superior camouflage. Antibiotic-resistant bacteria are favoured over susceptible bacteria, where antibiotics are present. But natural selection does not only operate on organisms. It operates on any entities that fulfil the appropriate conditions. If the entities exist in more than one form, if one form reproduces faster than the other, and if the 'offspring' resemble the 'parents', then natural selection is working. The conditions are often met by whole organisms, but they can also be met by cells, or molecules within cells.

The conflict in the ancestral eukaryotic cell was probably limited, as in the American–German business merger, because the two gene sets possessed complementary skills. The ability to feed on sunlight, or burn fuels in oxygen, was coded for by only one of the two sets of genes. Some skills would be lost if either gene set was completely

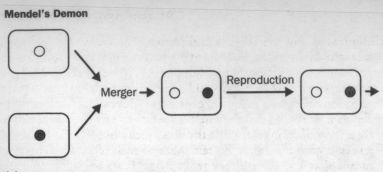

(a)

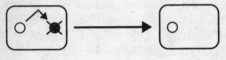

(b)

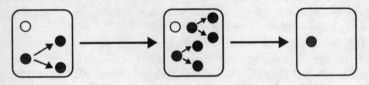

(c)

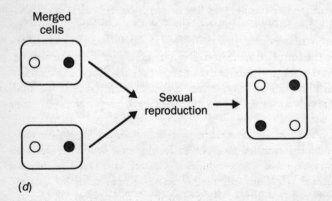

(d)

eliminated. But the skills would probably be coded for by only a minority of the genes, and one of the two sets of genes could replace a good chunk of the other without losing its useful skills. Also, the advantage of proliferating within the cell is so large that natural selection can favour a 'selfish' gene even if does some damage to the cell as a whole. (A selfish gene here refers to one that increases its own reproduction relative to the other genes in the cell.) If the cell has two gene sets, then the fair share of each is 50 per cent. This means that a gene on either set can double its frequency if it can somehow replace a gene in the other set with a copy of itself. With this kind of potential gain, the gene can afford to do some incidental damage. Indeed, the advantage of taking over the cell might exceed the benefit of the merger, and one DNA molecule would destroy the other completely. After a Darwinian flash of cloak-and-dagger genic assassination, or *vroom* of flat-out gene copying, the cell's affairs would be restored to their former state with only one of the two gene sets. A cooperative merger of the two gene sets, therefore, is not guaranteed. The successful eukaryotic merger requires some explanation.

The problem is compounded by sexual reproduction. Eukaryotic creatures tend to use sex. In a merged cell that breeds sexually, there are not two but four DNA molecules: two from the merger, doubled up because the cell has a father as well as a mother (Figure 8d). The conflict created by sex is potentially even more damaging than the conflict produced by the merger. Sex not only brings unrelated DNA molecules into the same cell, but also generates redundancy. The paternal and maternal genes do differ slightly, but they code for practically the same information. If a maternal gene kills the equivalent paternal gene, it doubles the chance that it will be passed on – and this kind of 'killing' is not just a theorist's fantasy, as we shall see. Sex creates a set-up like a business with two managers who

Figure 8. A merger creates a cell with more than one independently reproducing DNA molecule. (*a*) A simple merger with clonal reproduction produces a cell with two DNA molecules. Natural selection will favour whichever of the two can (*b*) destroy or (*c*) out-reproduce the other. If the merged cell uses sex (*d*), it will have four independently reproducing DNA molecules. Complex life needs cooperation among all the DNA molecules in the cell.

have identical jobs and abilities, and in which one manager can usurp the salary by taking over the other manager's job. One manager would now willingly murder the other. The business is damaged not only by loss of the dead manager's skills, but also by the unintended consequence of the act of murder itself. Bullets may ricochet off walls and hit innocent bystanders. There could also be a problem with the police and the law courts; but they do not apply to living systems, where the law of the jungle prevails. In some genetic examples, one gene does the equivalent of blowing up almost half the factory – the half where the other manager resides. The stakes are so high for the genes concerned that the welfare of the whole becomes a relatively minor concern.

Complex life would never have evolved if the eukaryotic cell failed to check the selfish genetic forces operating inside it. Any one destructive genetic act may have side-effects that reduce efficiency, just as in the business case. Worse still, natural selection exerts the same pressure on both the DNA molecules of a conflicting pair, and they could evolve to destroy each other. Then the cell really would be dead. If both managers blow up the other half of the factory, there will be no factory left. A subtler problem is that natural selection is less likely to favour increases in complexity if it can also favour genes that proliferate within the cell. Genes that devote themselves to intracellular punch-ups will not also contribute to the great creative arts of complexity. The problem is that the advantage of any increase in complexity will usually be small. There may, for instance, be an ecological opportunity for organisms that are more complex than the current members of their species, but the advantage would probably be only 1–2 per cent. Compare this with the double advantage gained by a mutant gene that abandons its proper function in the cell and instead evolves to scratch out its sister gene. Complex life required the genetic system to rule out the double-advantage selfish option, leaving the genes to explore the 1–2 per cent advantages of increased complexity.

The efficiency of a business can be compromised by innocent human error. Managers can make mistakes that harm the managers themselves as well as the business they are running. These mistakes are like the copying errors in the DNA that we thought about in Chapters 3–5. You can organize a business to reduce the rate of mistakes, and how much damage they do; but they will still happen, and may even ultimately limit the complexity that a business can attain. Our theme in the next three chapters will be a new kind of damage: damage that arises not from passive mistakes but from active selfishness. The efficiency of a business can be compromised by conflicts between its parts, conflicts in which one part gains by actively undermining another part, even though the efficiency of the whole business is reduced. Competition between businesses will tend to favour the businesses that suffer the least internal conflict. In life, natural selection can also favour selfish genes that increase their own copying rate but reduce the efficiency of the cell, or body, that they are in. These selfish genes are a new class of 'harmful' mutation. They are not just copying accidents, though they do originate as copying accidents, just as any new gene does. The copying accidents discussed in Chapters 3 to 5 were unambiguously harmful: harmful for the gene and harmful for the body. Natural selection simply worked against them. We are now concerned with mutant genes that natural selection favours at one level but opposes at another. The genes proliferate within the cell, or within one body, but harm the body as a whole. The evolution of complex life requires mechanisms to reduce, or purge, copying error. It also requires mechanisms that suppress or prevent internal conflict.

Genes that spread inside a cell are a particularly warped example of the theory that Richard Dawkins has made famous: the theory that natural selection favours genes that can produce the most copies of themselves. Dawkins describes genes as 'selfish', because each gene evolves to maximize the number of copies of itself, not of other genes. When we describe people as selfish we are saying something about their motives as well as the economic effects of their behaviour. The 'selfishness' of genes, however, refers only to their economic effects; genes do not have motives. During the evolution of complex life, the force of gene selfishness has had to be tamed. Complex life is built not only from many genes, but from many cooperating genes. The 60,000 or so genes of a human being all work together for the welfare and reproduction of the body they live in. The cooperation between the genes of a body did not just happen. It required special mechanisms for it to evolve, mechanisms that arrange affairs such that each gene is maximally selfish by being maximally cooperative with the other genes in its body. As we look at complex life today we are looking at the forms whose ancestors managed to resolve the conflicts among their genes.

A sexually reproducing, post-merger eukaryotic life form needed mechanisms to resolve conflicts between the maternal and paternal nuclear genes, between the maternal and paternal organelle genes (organelle is a general word for mitochondria and chloroplasts), and between the nuclear and the organelle genes. Chapter 7 will be about the evolution of cooperation in the nucleus. Here we shall assume that the nuclear genes are well behaved, and look at the relations between organelles, and between organelles and the nucleus.

The high noon of fertilization

The two cells that merged to form a eukaryotic cell probably contributed DNA molecules of roughly equal size (Figure 8). One of the DNA molecules has now evolved into the nuclear DNA, the other into organelle DNA. This evolutionary transformation has two main features. One is that the inheritance of the two molecules has evolved a sex difference. The nuclear DNA is reproduced sexually, and inherited through both parents. The organelle DNA is reproduced (or mainly reproduced) clonally and is inherited only from the mother – both sons and daughters inherit a clonal copy of their mother's mitochondrial DNA; the father's mitochondrial DNA is not (or mainly is not) passed on. The other feature is that the relative lengths and power of the two DNA molecules have changed, and the nuclear DNA is now far larger and controls the cell. The two molecules perhaps originally each made up 50 per cent of the DNA in the cell. In our cells, the DNA is (if we count nucleotides) $1/20,000$ mitochondrial and $19,999/20,000$ nuclear, or (if we count genes) $1/2000$ mitochondrial and $1999/2000$ nuclear. We shall see how the evolutionary changes in the size and inheritance of the two original DNA molecules have been two of the enabling mechanisms of complex life. We can start with the sex difference.

Some little-known algae are a good place to begin. The algae are a large group of living things that defy the simplest classifications of life. They photosynthesize, like plants, but they lack cellulose cell walls (which most plant cells possess) and some of them ingest food particles, like animals. Most algae are microscopic and single-celled and float in the water; but the best-known algae are sea-weeds, and seaweeds are macroscopic and many-celled. Algae matter for us because they reproduce sexually, with offspring formed from two parents, but some of them lack distinct males and females. The two parents look the same and they each produce a gametic cell of about the same size; the gametes fuse to form the offspring cell.* The new cell contains a nucleus, mitochondria and chloroplasts (as well as other cell contents) from each parent. The two nuclei soon fuse, but what happens to the organelles? The answer has been particularly clearly described for chloroplasts in the edible kind of seaweed called sea lettuce. In sea lettuce, and probably in most other isogamous algae, the chloroplasts from one of the parents are destroyed soon after the parental cells fuse. It is as if the maternal chloroplasts say to the paternal chloroplasts, 'this cell is not big enough for both of us.' A few minutes later, one set of chloroplasts is a pulped and swollen mess and the other set has the run of the cell.

The destruction of half the organelles is a remarkable inef-ficiency. A chloroplast is a complex machine that takes time and energy to manufacture. Why build a set of the things and inject them into the offspring if they are immediately going to be broken up and thrown away? Also, the damage is limited to the DNA-bearing organelles. Two complete sets of cellular apparatus are combined when the parental cells fuse: two nuclei, two sets of organelles and two sets of everything else. The two nuclei fuse, amicably enough, and everything else is retained – everything except for the mitochondria and chloroplasts. The mitochondria and chloroplasts from only one of the parents are tolerated in the offspring.

The normal rule in isogamous algae is the unilateral exter-mination of one parent's organelles; the sea lettuce illustrates this rule. There are hints of various exceptions, but we do not know enough about most of them to say more. One exception that does prove the rule is a microscopic aquatic creature called *Chla-mydomonas*. *Chlamydomonas* is an isogamous single-celled alga, in which two cells of equal size fuse to form a new offspring cell.

---

* The condition in which gametes of two parents are equal in size is called isogamy. Human beings, and most familiar forms of life, are anisogamous: the male gamete (the sperm) is much smaller than the female gamete (the egg).

The chloroplasts of both parents initially survive in the offspring and then fuse together a few hours after the new cell is formed. This looks like a peaceful exception; but we know a lot about breeding in *Chlamydomonas*, and the details show that it honours the rule in spirit if not letter. Just before the two chloroplasts fuse, the DNA in the chloroplast from one of the parents disappears. The full mechanism is known. A gene in the nucleus switches on soon after the new cell is formed, and releases an enzyme that smashes to bits the DNA inside one of the chloroplasts. *Chlamydomonas* probably has a more elegant substitute for the heavy-handed chloroplast destruction seen in other algae. Instead of destroying the whole chloroplast, it simply eliminates the small offending part of it – its DNA – and keeps and takes over the useful structure to perform photosynthesis.

These algae are peculiar in that both parents contribute organelles to the offspring. In most living creatures, the father ejects his organelles before the offspring cell is conceived. Sperm cells are small because almost everything except the nuclear DNA has been sucked out. But a few mitochondria from the sperm may make it into the egg, even in species with tiny sperm. Their fate has been observed, using the electron microscope, in mice. The mouse sperm mitochondria enter mouse eggs like missionaries walking into a cannibal feast. Here is a quote from an original description: 'Numerous bodies' gather round the sperm mitochondria, 'whereupon the mitochondria are degraded and digested.' In humans, almost a hundred sperm mitochondria may enter the egg, where they suffer some similar fate. The equipment used for the degradation varies from species to species, but the result is always the same. The mitochondrial missionaries are silenced forever in the maternal cooking pot.

The common theme in all these stories – about sea lettuce, *Chlamydomonas* and mice – is that organelles from more than one parent are not tolerated in the offspring. Nuclear DNA from both parents – yes. Cell contents other than mitochondria and chloroplasts from two parents – yes, in some odd cases (such as sea lettuce). Even organelles from two parents can be tolerated if those of one parent have their DNA cleaned out. But organelles from two parents, and with their genes intact, never seem to survive together for long in a living cell.

When both maternal and paternal organelles are found in one

cell, they are like two managerial teams in a newly merged business – in the extreme case in which the two teams have identical sets of skills and are competing for a fixed pool of remuneration. The competitive forces are unusually ferocious, and will not cease until one of the two has been eliminated. Natural selection will favour mitochondria that can eliminate or out-reproduce other mitochondrial types, and it also favours mitochondria that can avoid being destroyed or out-reproduced. One type of mitochondrion might have had the ability to recognize other mitochondrial types and destroy them, and would have thrived for a while. Selection would then have favoured counter-mutations in the victims. The victims might stop displaying the molecules used in recognition, or otherwise neutralize the instruments of destruction. The genes in the mitochondria from one parent are potentially in a perpetual evolutionary conflict with the genes in the mitochondria from the other parent.

The conflict between maternal and paternal mitochondria is likely to damage the cell. Mitochondria may physically destroy each other and in the worst case the two types of mitochondria could eliminate each other, leaving the cell with no mitochondria. Also, genes that code for vital activities will be converted to code for offensive activities, reducing the efficiency of the cell. Mitochondrial genes code for machinery needed to burn fuel, and the mitochondrion will not burn fuel so well if its genes evolve new offensive activities. Or genes may simply be deleted. It takes time to copy DNA, and a mitochondrion can speed up its breeding by deleting some of its DNA. The fast-breeder mitochondrion may then proliferate, but again at the cost of compromising its efficiency as a furnace. The cell as a whole loses out.

The nuclear genes are therefore dragged into the conflict. Natural selection will favour a nuclear gene that can stop the mitochondrial conflict. Mitochondria are needed to burn fuel, chloroplasts are needed for photosynthesis; neither are needed for their intracellular vigilante skills. It would be best for the cell if the mitochondria confined themselves to fuel-burning. The nuclear genes are like a higher level of management – a supervisory board, perhaps. They have nothing to gain from a victory of one side or the other in the conflict between the two junior managerial teams, but they do have something to lose if the efficiency of the business as a whole is reduced. The supervisory board could interfere in a detailed way, to stop particular tricks. A nuclear gene could also suppress the particular mechanism that the selfish mitochondria were using. But this is only a short-term expedient. The basic advantage to a selfish mitochondrion remains. A mitochondrion can always gain from selfishly eliminating other mitochondrial types even if it does reduce the efficiency of the cell as a whole. If one mitochondrial dirty trick is cleaned up by the nuclear genetic supervisory board, another will be invented soon enough. More and more of the nuclear coding capacity will be consumed by the mitochondrial problem. The supervisory board will find its time being used up in micro-managing the conflicts between junior managers. A complex business, or complex life form, will not be created, or survive for long, while the root cause of the conflict remains. The supervisory board may start to think about radical solutions.

The origin of gender

Complex life needs efficient cooperation between all the genes of a cell, which is hardly likely to evolve while crucial cellular components are engaged in private disputes. What can a nuclear gene do about these eternally squabbling organelles? The evolutionary answer seems to have been – create gender.

Imagine an ancestral stage at which living creatures reproduced using cells – gametes – that were all the same size and all contained organelles. The gametes were like the gametes in modern sea lettuce and *Chlamydomonas*. When two such cells fused, their organelles fought it out and the cell was damaged. Now a mutant nuclear gene arises; it acts to keep its own organelles but exclude those of the other cell. The mutant will be rare when it first arises, which means that it is most likely to meet and fuse with a gamete of the majority type rather than another copy of itself. When it fuses with a standard gamete, it excludes all (or perhaps most of) that gamete's organelles. The 'organelle-excluder' mutant gene then finds itself in offspring in which there is no, or reduced, conflict; it has prevented one of the conflicting parties from being present. Natural selection favours the gene when it is rare and it spreads. Another kind of mutant nuclear gene can also spread when rare. It is an 'organelle-ejector' mutant, which ejects its own organelles before fertilizing another cell. When it first arises it will be rare and will usually fuse with a gamete of the common type. It will tend to find itself in cells with reduced conflict. Natural selection favours it and it increases in frequency. But as it becomes common, it becomes more likely to fuse with another gamete like itself. They have both ejected their organelles and fertilize each other to produce a cell with no organelles at all. This is unsatisfactory, and natural selection will stop favouring the organelle-ejector gene. The organelle-excluder type faces a similar difficulty as it increases in frequency. But things are perfect when an 'organelle-excluder' fuses with an 'organelle-ejector': they complement each other like soulmates in a romance. When either kind of gene (such as the organelle-ejectors) have spread to a decent frequency, they create an advantage to the other kind of gene (organelle-excluders). Natural selection will now act to make the gametes fuse only with gametes of the complementary types. The population has then evolved from an initial state in which all the gametes were the same, to a state with two kinds of gametes, one of which ejects its DNA-bearing organelles before reproduction and the other of which retains them and excludes its partner's organelles.

The way that natural selection works on the 'ejector' nuclear genes is a striking feature of the theory. Selection is favouring a gene in the nucleus that evicts organelles from its own cell. We might suppose that, in the conflict between organelles, the nuclear genes would support their 'own' organelles – the organelles inherited from the same cell as the nuclear genes came from. Not a bit of it. There is no evolutionary common interest between genes in the nucleus and genes in the mitochondria. The ejector genes do what nuclear genes now do in males. Male nuclear genes evolutionarily need to end up in a cell with good mitochondria – mitochondria that burn fuel well – but it does not matter whom these mitochondria come from. A gene in a male nucleus can ensure that it ends up in trouble-free offspring by banishing mitochondria (as much as possible) from the sperm that are to carry the nuclear genes into the next generation.

Gender is one of the enabling mechanisms of complex life on Earth. It did not evolve in order to allow the future evolution of complex life, but it did make this evolution possible. The complexity of life would have been limited to a much lower level if life had failed to solve the problem of organelle conflict. Gender solved the problem by producing offspring cells that were trouble-free. The mechanism by which gender resolves genetic conflict is one of the two great conflict-resolving mechanisms that we shall meet in this and the next chapter. Gender makes the organelles in a cell genetically uniform, and this prevents natural selection from acting among them. In a fertilized egg that contains both maternal and paternal organelles, there is more than one kind of mitochondrion. It is this variation, within the cell, that fuels natural selection. Genetic selfishness is an automatic consequence of natural selection, and one way to put a stop to it is to stop natural selection from operating. It can be stopped, at least locally, by removing all the variation. Natural selection cannot do anything with a uniform population. Genetic uniformity undermines genetic selfishness. The future evolution of organelles is then channelled into creative, rather than destructive, lines – and it is no coincidence that the highest achievements of complexity on Earth are all in life forms that are gendered, with separate males and females.

The losers in this tidy deal are the genes in the male mitochondria. They will not, in an evolutionary sense, have taken it lying down. A male body is one big mitochondrial graveyard, and the mitochondria will do everything in their power to avoid entering one. Mitochondria unlucky enough to end up in a male body will try to redirect it away from its disgusting masculine reproductive habits. I like to imagine that the mitochondria in an egg, before it is fertilized, will take an interest in whether the cell is fertilized by an X- or a Y-bearing sperm (X and Y stand for the chromosomes that determine sex: if the egg is fertilized by an X chromosome it will grow up female, and if by a Y it will grow up male). The mitochondria ought to line up behind the membrane of the egg, as if with their noses pressed against the window, viewing the aquarium of sperm racing outside. When they see an X-bearing sperm swim towards their cell they let out a great cheer of encouragement; when they see a Y-bearing sperm they boo and make rude signs at it. In us, these efforts have little effect because the sex ratio is about 50 : 50. A similar process produces more dramatic effects in many hermaphrodite plants. In these plants an individual should be able to breed both as a male and a female, but many of them can breed only as females because the male reproductive system has been internally switched off. Male reproduction has been switched off by the action of organelle genes.

The organelles are not inherited through the pollen, and male effort is (for them) wasted effort. The organelles do better if the plant concentrates on producing the eggs that can carry them forward. Some delightful facts are known about organelle genes that repress male reproduction and nuclear genes that restore it or repress the repressors. Research so far has probably only scraped the surface. The nuclear and organelle genes are trapped in an eternal Freudian conflict, as natural selection on the organelle genes will cause them to smother each new attempt by the nuclear genes to assert gender equality.

Leda Cosmides and John Tooby are two of the inventors of the organelle theory of gender, and they pointed out that it explains certain facts about mitochondrial numbers in the egg. I described the female gamete as simply excluding other cell's organelles and retaining its own, but this is not the end of the matter. In a conflict between two organelle types, numbers count: if there are more of one organelle type, it will be more likely to survive any conflict and be passed on to the daughter cells when its cell divides. An egg is a big cell, which should help because a big cell can house large numbers of organelles; but consider now how the egg comes to be big. The reproductive cells are produced in a special cell division called meiosis. Meiosis actually consists of two rounds of cell division, and one cell has become four cells by the time it is complete. All four of the products of meiosis become sperm cells in males. But in females only one of the four becomes an egg; the other three cells get wasted on the way. The sex difference makes sense if selection has acted to multiply the numbers of organelles in the egg. The cell divisions in egg manufacture are unequal: one of the cells is filled with good organelles while the other is sent away empty. It is of course crucial that the full, rather than the empty, cell should become the egg – as it indeed does. The meiotic cell divisions that generate every egg are unequal in this way, whereas the sperm-generating meiotic divisions of males are comparatively equal. This modern sex difference follows from a past history of selection to build up the mitochondrial forces in eggs.

Egg cells in fact are exceptionally full of mitochondria. A typical mammalian cell might have about twenty or more mitochondria. The numbers vary hugely – muscle cells have heaps of mitochondria while skin cells almost do without. The cells in the line of cells that leads to an egg have 100–200 mitochondria up to the last (or last two) cell divisions in the line. The numbers are then banged up to around 100,000 before the cell finally matures into a gamete. This is theoretically satisfying. It again appears that the mitochondrial forces are peculiarly massed in the egg. Also, the policy of keeping the numbers down and then cloning them up hugely later on results in more genetic uniformity among the egg mitochondria than if the increase were done some other way. The mitochondrial numbers might instead have been raised by summoning them from many cells or by multiplying the numbers earlier in the cell line, but the mitochondria in the egg would then be more likely to show genetic differences. The mitochondria that enter the offspring are more likely to be genetically identical if the multiplication event is left to the last minute and concentrated on a smallish number of lucky founders.

There are many more facts that fit the theory. For instance, some species, such as pine trees, seem to implement the deep strategic imperative of the theory – organelle uniformity – by the reverse mechanism: the mother dumps her organelles and the father passes his on. Other creatures, mainly obscure microbes, lack gender. They are unisex creatures, without distinct males and females. They have the kind of 'recreational' sex that we looked at in Chapter 5. Two cells meet, swap copies of their nuclei, and separate. They do not exchange organelles, and there is no danger of a postcoital mitochondrial row. The evolutionary condition for gender is absent, and gender is absent. As a final detail, the missionary sperm mitochondria of mice are labelled for destruction in the egg by a molecule coded for by a male nuclear gene. Before the sperm are sent into the female reproductive tubes, the male nucleus effectively paints on each of the sperm mitochondria EAT ME PLEASE. Then the egg knows which mitochondria to pick on. The male nuclear genes have evolutionarily ganged up with the female egg to ensure that the male mitochondria are not a nuisance.

Gender seems to have evolved to impose cooperation on mitochondria. Living bodies go to great lengths, by a variety of mechanisms, to ensure that the mitochondria in a body are genetically uniform and to reimpose uniformity if it is spoiled by intruders. We have now solved the first of the two problems that we started out with. The problem was to understand the sex difference in the inheritance of organelles. The solution is that it prevents a conflict from taking off between organelles in the offspring. The eukaryotic cell merger created the potential for more complex life, as it brought together the genetic information of two cells into a single cell. The evolution of the sex difference in organelle inheritance was one of the two or three main post-merger conflict-resolving steps that allowed the extra genes of the merged cell to be put to use creatively, rather than destructively, in the evolution of complexity.

Imagining gender

If you asked people of various backgrounds what they thought was the deep difference between males and females, you would receive many replies. Some would deny there is a deep difference. Some would psychobabble about a nurturant sex and a competitive sex. Geneticists would point to the XX and XY chromosomes, economists to the relative parental investment in offspring, veterinary students (and others who go around peering between animals' legs) to the shape of the creature's genitals. Freudians, fundamentalists and sexual politicians could add their own ideas. But from the perspective of organelle conflict, all these criteria (in so far as they are valid) are superficial. They specify things that derive from the deepest of the sex differences. Sex chromosomes, genital shape, reproductive investment and perhaps even personality all arguably follow from the resolved conflict of the organelles. The deep nature of maleness is to eject organelles from reproductive cells; the deep nature of femaleness is to keep them.

When natural selection favoured life forms with a pure stock of organelles, uniform and free of troublemakers, it split the members of every sexual species into two complementary types: organelle-keepers (that is, organelle-excluders) and organelle-ejectors. The gametes of the organelle-ejecting type will be small, and of the organelle-keeping type large: they will evolve into sperm and eggs. Many other differences between males and females will then evolve after this basic division into fat gametes and thin gametes. A male can produce more sperm than a female can eggs, and access to the reproductive systems of females therefore limits the number of offspring a male can produce. Female reproductive output is not so limited by their access to males. Selection favours anything in males that enhances their attractiveness or competitive power in the mating marketplace; in females it favours an ability to distinguish high-quality from low-quality males. Darwin first invented this theory, and he called it sexual selection. It can explain much of natural history – that is, sexual natural history such as the peacock's tail and the narwhal's tusk. It predicts a general pattern of male competition and female choice, though many other particular factors operate in particular species. Human beings, for instance, use pair bonds, which set up a selection pressure for female, as well as male, competition, and for male, as well as female, choice. Not everyone accepts an evolutionary explanation for human sex differences, but at the very least evolution explains why human society is gendered at all, as opposed to ungendered.

The theory of gender that we have been looking at here has an interest on a much grander scale than the curiosities of pairing in any one species. The theory implies that the existence of gender itself, in life on Earth, is an evolutionary fluke. Gender follows from sexual reproduction, but is not logically required by it. Sex does require two individuals to mix their genes, but the two individuals do not have to differ in any other respect. They could have unisex genitals, unisex wiring and plumbing, unisex bodies and unisex personalities. Gender evolved to stop the offspring from inheriting warring sets of organelles, from two parents: but this problem only ever arose because sex had evolved in a eukaryotic organism, with a merger in its ancestry. If complex life had evolved without a cellular merger, the cells of complex life forms would not contain organelles that bear DNA and reproduce independently, and gender would have been unnecessary.

To be more exact, gender depends on two flukes, or events that look like flukes. One is that complex life evolved via the merger; the second that both the DNA sets of the merger partners were retained in the post-merger cell, rather than one being completely eliminated or incorporated into the other. Complex life probably could have evolved without a merger. It could have evolved by increases in the gene numbers of a single life form, with the increases evolving from the standard copying accidents in DNA that we met in Chapter 2. Alternatively, complex life could have evolved via a merger, but the two DNA molecules could have been rapidly reduced to one. Either way, the increasingly complex creatures would eventually have hit a limit to gene numbers, set by the mutation rate. Natural selection would then have favoured sex, and sex would have evolved. But the sexually reproduced offspring would not contain any DNA-bearing organelles; all the DNA would be in the nucleus. Natural selection would not favour gender. Reproduction would happen when any two individuals in the population mixed their genes in an offspring.

I do not know for sure that the two flukes really were flukes. Bacteria may for some reason be unable to evolve increased complexity by gradual steps, within one species. In this case, complexity could evolve upwards only after something like the revolutionary step of merging two bacterial cells. Without the successful merger, life on Earth would have continued indefinitely in the form of simple bacteria, rather like it did for the first half of the history of life. Complex life would evolve either via a merger or not at all, and our origin in a merger event would be a retrospective inevitability, not a fluke. Our understanding of why some genes have been retained in the organelles is also inadequate. We may plausibly argue that the genes could all have been moved to the nucleus or been lost, but the argument may be wrong. There may be some reason why some genes have to be in the organelles. It would be inevitable that the post-merger cell would retain two separate sets of DNA. My judgement, however, is that complex life did not have to evolve via a merger, and the post-merger organelles did not have to retain their genes. They were two of evolution's unnecessary accidents, and anything that has followed from them is, in the cosmic view, also accidental.

Complex life elsewhere in the universe will usually lack gender. It will probably use sex, because copying error is probably universal; but it will usually not have separate males and females. It is a wonderful science-fiction question, to ask how evolution would work out in this case. What would life be like if it were as complex as us, but lacked gender? We are obsessed with gendered relationships, and without gender the themes of television programmes, the appearance of magazine racks, the innumerable gendered details of human art, human commerce, human society, even human thought would all be transformed. Pair bonds could still exist, but with some different imaginative form. Perhaps those pair-bonded couples would imagine themselves less as a *yin* and *yang* of complementary but different types, and more as a pair of interchangeable individuals who can produce more as a pair than singly – like two loudspeakers in a stereo system.

There would still be mating market forces. There would be much the same advantage to breeding with a partner of high genetic quality as there is in a gendered species. The life form would have evolved to display genetic quality by some procedure, analogous to frog croaks, but everyone would display their genetic quality and assess the quality of everyone else. High-quality individuals would be competed for as mates, and the law of supply and demand would assert itself. The difference from us is that an individual can now pick its mate from the whole of the rest of the population, rather than the half of it that is of the opposite sex. You may think this is an improvement, because the number of potential partners will be doubled, but so too will the number of rivals. There will be no place to hide. There will be no single-sex changing rooms, clubs or luncheon dates. The all-intruding mating market will extend into every corner of social life. The doubled scope for mate choice might mean that the genetically superior individuals would enjoy rather more of a reproductive advantage, and produce more of the children, than in our gendered system. The more powerful, more invasive mating market would then permit life to evolve to be more complex than here on Earth. I should also not be surprised if the individuals with a set of good genes were the sucked-up-to 'top cats' and arbiters of taste. In that other-worldly, no-merger, gender-free place, social and family life might be transformed, but consumer products might still be endorsed and political causes promoted by the (*mutatis mutandis*) smiling faces of the fortunately mutation-free members of a unisex population.

The concealed workings of the master artist

We can draw a general moral that we shall meet more than once as we think about complex life. The grand structure of the theory we have been looking at is as follows. We first showed that there is a conflict between genes in paternal organelles and genes in maternal organelles, and between nuclear genes and organelle genes. We then used the theoretical conflict to explain a deep feature of life: the male–female phenomenon. We did see some examples of overt conflict, such as those missionary sperm organelles that were butchered in the egg. But the overt conflict is difficult to find, and you need to know where to look. You can find it in some isogamous, single-celled algae: but these are microscopic creatures that hardly anyone has heard of, and in any case most single-celled algae are not isogamous, and even isogamous algae have methods of resolving organelle conflict. You can also find it in mice, but it is a fleeting business, concentrated in a few hours after fertilization and concerning only a few mitochondria. If the theory mattered only to explain these observable kinds of conflict, it would be trivial. But this is not where the importance of the theory lies. Its importance is in explaining big facts of life that would otherwise not make sense. Here it explains maleness and femaleness.

Once males and females have evolved, the conditions for an observable conflict between organelles have been removed. We no longer expect to see a conflict. The theory still matters, because it explains the existence of a mechanism that prevents the conflict from happening. Let me give a social analogy. Society contains some people who own desirable property and wish to keep hold of it, and other people who do not own such property and would like to take it from the people who do. There is a conflict of interests. Society also contains mechanisms for dealing with the conflict, such as the police force and the law courts. A society could in theory exist in which the police and courts were perfectly efficient at catching and deterring thieves. In this society you would see no manifest conflict between haves and have-nots because the conflict has all been prevented. You still need to know about the conflict, however, to understand the society, because if you do not know about it the police will be an existential puzzle. If you do not know about organelle conflict, gender poses an analogous puzzle. In fact it acts to police a genetic conflict. The conflict resolution is so successful that now, after the mitochondria are squeezed out of the sperm, or sent into the trigger-happy egg with targets painted on them, the creative force at work is invisible. The mark of a great conflict-resolution device is that a naive observer would overlook not only its existence, but even the need for it to exist. In the greatest art, you do not notice the artistry.

The theory is also important because the conflict-resolution device is a precondition for the future evolution of complexity. In the evolution of complex life, natural selection has to act to stop itself from acting within a body. When natural selection favours nuclear genes that cast out the organelles from the gametes, it stops itself from working on the organelles in the offspring's body. The organelles are then forced to cooperate and make the whole body more efficient. The organization of, and cooperation between, parts within a body contrasts with the relations between individual bodies in a species. Human society does possess supra-individual organizational properties, including the police and legal systems, but they are clumsy and inefficient. Social cooperation does not begin to compare with the harmonious relations between the different parts inside our bodies. Without gender, the complexity of living organization at the level of individual bodies would be limited to something analogous to the social level of organization that we see in life now.

The Snark was a Boojum

I now want to turn to the second big fact about the post-merger eukaryotic cell. The cell probably began with two DNA molecules of roughly equal size. This has evolved to a state in which the nuclear DNA is much larger, and dominates the cell, and the organelle DNA has shrunk into near-vestigiality. The change has been partly achieved by gene transfers, as can be deduced from a careful inspection of the genes in the nuclear DNA. Some of them clearly have a mitochondrial origin (it is as if an old library stamp revealed the origin of a book). We do not know how much of the gene reduction from the ancestral mitochondria has been caused by simple loss and how much by migration to the nucleus, but both processes have contributed.

We can see something of how the transfers have happened by looking at the mitochondria in a variety of modern life forms. The mitochondrial DNA of all modern animals has the same small size, but it is larger in some single-celled creatures. A single-celled microbe that lives in fresh water has the largest mitochondrial DNA molecule found so far. The creature is called *Reclinomonas americana* and its mitochondria contain ninety-seven genes. Mitochondrial DNA molecules have a range of sizes in modern life. The range could in theory be due to separate origins: it could be that the modern mitochondria with large DNA are descended from one ancestor, with large DNA, and the modern mitochondria with small DNA from another ancestor, with small DNA. In fact, however, all mitochondria, in all eukaryotes – in all plants, all animals, all fungi and all the single-celled eukaryotes – are descended from one origination event: there is no evidence for multiple origins. The variety of modern mitochondria instead represents several stages in the exodus of their genes. *Reclinomonas americana*'s mitochondrial DNA is relatively 'bacterial', but it has still lost about 90 per cent of its ancestral gene set – its DNA is only 10 per cent of the size of the DNA in even a small bacterium. Other single-celled eukaryotes have smaller mitochondria still, and those of animals (with thirty-seven genes) have shrunk by another two-thirds below the *R. americana* stage. The modern variety of mitochondria shows that the mitochondrial genes have softly vanished away in several evolutionary stages. The chloroplasts tell a similar story.

The evolutionary migration of genes from one ancestral DNA molecule (now the mitochondrion) to the other (now the nucleus) makes sense for several reasons. One is the familiar advantage of combining all the genes in a cell in one reproductive unit: it makes it more likely that the genes will all cooperate. The paternal and maternal mitochondria are particularly unlikely to cooperate because they are independently reproducing units and their gene sets are identical: a cell does not need two sets of identical mitochondrial genes. Gender reduces the two reproducing units to one by the drastic procedure of eliminating the paternal mitochondria. The same basic problem existed in the ancestral eukaryotic cell just after the merger (Figure 8). The two gene sets were not identical: indeed they were partly complementary, and this would have reduced the conflict. But the potential for conflict was still there. The conflict could be stopped not by eliminating one of the gene sets, but by combining them into one. The unification is incomplete, because the organelles retain some DNA; but most of the genes have been incorporated in a single reproducing unit. The shrinking of one DNA molecule and the expansion of the other was another crucial step in the evolution of complexity. It reduced the scope for conflict between DNA molecules in the same cell.

The gene transfers have another advantage. In the early conflict between two DNA molecules, there would probably have been a 'winner' and a 'loser'. Genes in one of the sets might, at least temporarily, have been ahead in the conflict – able either to damage the other set (Figure 8b) or to out-reproduce them (Figure 8c). In the case of damaging conflicts, genes in the 'losing' DNA can benefit by joining the victors. A gene incorporated in the victorious DNA will escape being damaged and excluded from daughter cells. The process would snowball, because the DNA molecule that grows (over evolutionary time) also becomes increasingly likely to win in a conflict of destruction. If the larger DNA molecule is damaged, the cell suffers more than if the smaller DNA is damaged; the smaller DNA has more need of the larger DNA than vice versa. Also, the larger DNA is more likely to evolve superior armaments, because it contains more genes. The more genes a DNA molecule contains, the more kinds of weaponry and defence it can evolve. In the evolutionary gene transfer, as Jesus put it, 'unto every one who hath shall be given, and he shall have abundance: but from him that hath not shall be taken away even that which he hath.'

Something similar happens if natural selection is working not on destructive power, but on the rate of reproduction of the two DNA molecules. The words 'winner' and 'loser' take on an ironic meaning in this case. We might think of the DNA molecule that transfers its genes, shrinks and is humbled within the cell as the 'loser'. This is indeed the standard interpretation of the historical events. But natural selection positively favours the DNA molecule that transfers away its genes, not as some kind of defeat but as a way of gaining an advantage. As I have said before, a DNA molecule can accelerate its reproduction by deleting part of itself, simply because DNA takes time to copy. Natural selection could then favour mitochondria that dispose of their genes. The details of the disposal would probably depend on the genes. The merged cell contains two full sets of genes. Some of the genes in each set will do much the same as a gene in the other set, whereas other genes will be unique and code for useful skills in the cell. For the genes in double copy, the first of the DNA molecules to dump its copy would probably gain an advantage – immediately because of faster copying and in the medium term strategically because the other DNA cannot dump its copy (if it did, it would kill the cell). A disposal of such a gene can be brought about by a 'deletion' mutation. The unique genes, however, cannot simply be deleted. They require another kind of mutation in which the gene is excised from one location and transferred to another. (Mutations of this kind are well known to take place; geneticists call them translocations.) The molecule that loses the gene is not losing evolutionarily: it will now copy itself faster and slow down the copying of the other DNA molecule. Natural selection provides an active shrinking force. Modern mitochondria look as if they have been through a history in which they have stripped down for rapid breeding. Mitochondrial DNA is mean as well as lean. It has no junk DNA, nor any of the other luxuries that nuclear DNA allows itself. Junk DNA would have slowed down the copying of mitochondrial DNA, and natural selection has purged it.

It may therefore be naive to think of the nuclear DNA as having 'won' and the mitochondrial DNA as having 'lost'. It is true that, of the two original DNA sets in the merged cell, one has enlarged and become dominant while the other has shrunk. The mitochondria have almost lost their reproductive independence. All the genes concerned with copying mitochondrial DNA have moved to the nucleus; the only genes remaining in the mitochondrion are used in the most proletarian functions. The timing and extent of mitochondrial reproduction are all now under nuclear control. Surveying these facts, it is easy to conclude that the nucleus has 'won' the evolutionary conflict. But the true measure of victory, in the modern theory of evolution, lies in the quantity of genetic information on Earth that has one form rather than another. Who actually copies this genetic information, to result in its abundance, is a secondary matter. If the genes that were formerly in one DNA molecule are now copied as part of the other, the quantity of those genes is not reduced. The mitochondrion need not sink into vulgar argument when the nucleus declares a nuclear victory after each new gene transfer. It can quietly sit back, tap its snuffbox and watch with amusement as the nuclear assembly lines toil and sweat to fill the world with mitochondrial genetic exports.

Experiencing natural selection

The kind of natural selection that I have been discussing – selection on mitochondrial DNA inside the body – is not just hypothetical, or confined to the past. It is probably going on inside the cells of all of us now, and in a vitally important form. It is probably part of the reason for ageing. Any one of our cells contains many mitochondria: ten to a hundred or so, depending on the cell type. If one of these copies itself in greater numbers it will proliferate in the cell and be more likely to be inherited in the daughter cell. Natural selection favours the mitochondrial fast-breeders. It may seem pointless or even evil, but apparent pointlessness and evil never put a stop to natural selection. Natural selection just happens, when the conditions for it to operate are present. A mitochondrion may gain a reproductive advantage in many ways, and deleting part of its DNA is almost certainly one of them. We can predict that mitochondria with deletions will occasionally spread in the cells of a body. And it appears that they indeed do. A human being starts life with a good set of mitochondria. Later, a 'selfish' mitochondrial type may take off in some unpredictable subregion of the body. One study looked at a mutant mitochondrial type that had 5000 letters deleted. Normal human mitochondrial DNA is 16,000 letters long; the DNA of these mutants was 11,000 letters long. The study measured the frequency of the mutant mitochondria in a sample of human beings of various ages. The mutant mitochondria had a frequency of zero per cent in newborn human beings, but it rose to a detectable fraction in the skeletal muscle of people aged twenty years or more. Muscular performance – like everything else in our bodies – deteriorates with age, and the spread of selfish, autodeleted mitochondria is probably part of the cause. A mitochondrion that has deleted a third of its DNA will not work so well, and the muscle that it occupies will be relatively short of energy. Wrinklies puff and wheeze where youth does effortlessly tread.

Mitochondria that contain deletions in the DNA have also been implicated in specific illnesses. Cardiac ischaemia is a disease in which the blood supply to the heart is inadequate, and one study of the heart muscle of people suffering from chronic cardiac ischaemia found that mitochondria lacking a 5000-letter region in their DNA were 8–22,000 times more common than in normal healthy people. One sixteen-year-old girl died of heart disease and her heart muscle happened to be studied by mitochondrial DNA experts. The muscle was stuffed with mitochondria containing a 7400-letter deletion. These stories could be the tip of an iceberg. However, I should stress that although defective mitochondria accumulate with age in our bodies, it is not known that they proliferate because of the kind of natural selection we have been thinking about here. Some kinds of genetically shrunken mito-chondria do copy themselves faster than normal mitochondria, but we do not know whether the disease-causing mitochondria do so too. The idea is plausible, but research is needed to test it.

Mitochondrial DNA may simply contribute to ageing and disease roughly in proportion to its importance in the body, but there are suggestions that it contributes disproportionately. For instance, mitochondria use only thirteen proteins that are coded in the mitochondrial DNA and more than a thousand proteins that are coded in the nuclear DNA. We might expect that about 99 per cent of genetic diseases affecting the mitochondria would be due to defective nuclear genes, and 1 per cent to defective mitochondrial genes. In fact almost all mitochondrial genetic dis-eases are due to defects in the mitochondrial DNA. One explan-ation is the peculiar force of selection on mitochondrial DNA, in favour of deletions or other tricks that accelerate mitochondrial reproduction in the body. If so, our accidental origin in a eukaryotic cell merger is one more dark hour in our evolutionary ancestry. We should enjoy better health and a more vigorous old age if our mitochondria had evolved by some other means, or if a complete series of gene transfers had cleaned all the DNA out of our dispositionally selfish mitochondria.

We can make sense of one other fact about mitochondrial diseases. They are almost always found in only one part of the body: most of the body's cells are normal, but one subregion, such as heart muscle, goes bad. The diseased subregion is where a mutation has taken place within the body of the individual. We have seen several mechanisms that ensure that our bodies begin life with a uniform set of mitochondria. Our genetic uniformity at conception helps to stop natural selection from acting within our bodies, but mutations during our lifetimes produce the raw material for it to start again. Our bodies mature into an internal genetic jungle and we experience a cruelly Darwinian limit to the complexity of our lives.

The mysterious eleventh-hour Snark-deterrent

The forces, and the benefits, of genetic aggregation look so power-ful that it is something of a mystery why there are any genes left in the organelles. It may have something to do with local control or (in Eurospeak) genetic subsidiarity; but no one has worked this idea out in convincing detail. Or it may have something to do with the kinds of protein that can move across the mitochondrial membranes – but no one has made this theory convincing either. The most popular positive idea is that evolutionary gene movements have been limited by the peculiar genetic code used in mitochondria. The genetic code is the relation between DNA language (the four bases) and the amino acid building blocks of proteins. The genetic code is often said to be universal – and to a good approximation it really is universal – but some mitochondria use slight variants of the genetic code.

A change in the genetic code will make gene transfers more difficult, for much the same reason that human beings have difficulties when going to live in a country where they speak a different language. It may be better to stay at home, where people understand you. However, this idea is also unconvincing. Not all mitochondria have deviant genetic codes. We do, and so do other animals. But other creatures use the standard genetic code in the mitochondrial genes, and a change in the code cannot be the reason why they have retained genes in their mitochondria. Perhaps the explanation can be made to work in a more complicated, two-stage version. On the limited evidence we have, the code seems to have changed only in species with drastically reduced mitochondrial genomes. It may become easier for the code to 'drift' after the mitochondrial DNA has shrunk to only a few genes. The code would be constrained to remain standard while there were many genes. But as these genes transferred to the nucleus, the code would change, preventing the final transfer of all the genes. Maybe, or maybe not: we do not really understand the exceptional mitochondrial stay-at-homes. If we did, we might have a better understanding of gender and ageing. Would some future eukaryote that managed to dispose of its remaining organelle genes be on the evolutionary road to a reduced-senescence, gender-free politics of reproduction?

However that may be, the stay-at-homes are only exceptions – minor exceptions – in the big picture. The big fact is that about 98 per cent of the original mitochondrial gene set has been lost or transferred to the nucleus. This makes sense, as we have seen, given the way natural selection works in a cell with two DNA molecules. The relative sizes of the modern nucleus and mitochondrion, however, are not only due to gene transfer and gene loss. Since the original eukaryotic merger, the nucleus has actively increased the number of its genes, in addition to the genes it has acquired from the mitochondrion. The original eukaryotic cell might have had about 5000 genes; but modern human nuclear DNA has some 60,000 genes. The factors we have already looked at explain why mitochondrial genes transfer to the nucleus. They also implicitly help to explain why the nucleus rather than the mitochondrion later expanded its gene set. Here we can add a third factor, one that combines the theory of gender in this chapter with the theory of sex in Chapter 5. The theory of gender gives a good reason why mitochondria have to be copied clonally, because they have to be inherited from only one of the two parents. The theory of sex gives a good reason why the size of clonally copied DNA molecules is limited: if they expand, the number of copying errors goes up and the DNA will eventually undergo a mutational meltdown. Modern mitochondria have small DNA, comfortably below the mutational danger zone, but the ancestral mitochondrial DNA could never have expanded far before it hit the limit. Nothing stopped the nuclear DNA from evolving sex, and it was free to accumulate the extra genetic information as more complex life forms evolved. The organelle-excluding *yin* gene and the organelle-ejecting *yang* gene originally evolved to prevent destructive squabbles among organelles. But they also fixed the future pattern of gene increases. The evolution of gender locked the organelles into eternal virgin birth, and sexual reproduction has become an additional cause of the expanding disparity between nuclear and mitochondrial DNA.

Catch-22

If some of the arguments in this chapter have seemed a little contorted, we can finish by finding a positive merit in these contortions. The merger of the original eukaryotic cell unleashed a horrid pack of evolutionary forces, which have worked themselves out in some messy and devious ways. But the results may help to explain the possible evolutionary 'difficulty' of the eukaryotic cell. In earlier chapters, I have argued that the evolution of the eukaryotic cell was exceptionally improbable and that mergers may be rare relative to other mechanisms by which gene numbers increase. The eukaryotic cell may be the only merger in the history of our 60,000 genes, accounting for only 1 or 2 per cent of our gene set. There could have been some other mergers in the early, anarchic stages of life, and the contribution might be a bit higher; but I doubt whether mergers have contributed more than 5 per cent of our genes. Even on a generous estimate, mergers have contributed only a few of our genes. What is so difficult about them?

One answer is that it is evolutionarily hard to make two independently replicating DNA molecules cooperate inside one cell. Natural selection will tend to work in exactly the opposite direction. The initial set-up was like a parasite inside a host (indeed, the initial set-up may well have been a parasitic cell inside a host cell). The two DNA molecules were competing to use the cell to replicate themselves. The Darwinian forces are antagonistic – tending towards exploiting the cell for competition, not pooling interests. Soon the only genes that would be left would be the greediest or most aggressive, and any surplus DNA would be auto-excised and lost. The life form is likely to collapse back to something like its pre-merger state before any increase in complexity can be established. Genes cooperate only if there are mechanisms to enforce it. Mechanisms of this sort exist if gene numbers increase within one DNA molecule. But if gene numbers increase by a merger, no such mechanism exists and they have to be evolved from scratch. The genes of one merger partner will not just combine with those of the other and copy themselves in a unified, orderly way. The two sets of genes will have different replicating systems that have been evolving separately for millions of years. The obvious result of forcing them together will be chaos. It will take time to build, by gene migrations or gender or some other mechanism, the political infrastructure that might enable the two to work and replicate together. The merger could not lead to increased complexity before that infrastructure had been built, but the building takes so long that the extra genes would probably be lost before it was in place. The raw materials of complex life would have evaporated before anything creative could be done with them.

I suspect that mergers and acquisitions are more frequent in the world of business than of life, and more frequently lead to long-term increases in complexity. Businesses are better at mergers and acquisitions than life is. The reason is that businesses have two mechanisms that life forms have failed to evolve. First, businesses employ strategic planners, who are on the look-out for acquisitions that will improve the competitiveness of the business. If a business knows it needs some software-writing skill, it can either try to develop it in-house, or it can scout around for people who can do it, and buy them. It is as if a life form had, in its cognitive system, the ability to detect which other life forms could be beneficially merged with. Human beings now, for instance, might benefit from some chimp expertise in the immune system, to resist AIDS. But even in this case we do not know exactly which DNA codes we want from them. In so far as we have any knowledge of desirable (partial) merger partners, it is because of science; all other life forms completely lack the skill that every large business possesses. Second, businesses can integrate the command structures after a merger. When two management teams come together, there is the same danger as in a life form with two replicating DNA molecules. But the integration of the eukaryotic DNA to form one unit (or one and a bit) from the two ancestral units took a long time and depended on genetic accidents – accidents in which genes were transferred between DNA molecules. The integration of managerial command structures after a business merger is enacted in a ruthlessly directed way, and leaves nothing to accident. It is as if our cells contained mechanisms that could recognize useful codes in the DNA of another species, excise them from their surroundings and incorporate them into our regular DNA. The rest of the codes from the other species would be disposed of. We contain no such mechanism.

Symbiosis in life is not uncommon, though the two symbiotic species usually retain their identities and combine in a limited way only, for temporary advantage. Most symbiosis is more like contractual cooperation between two businesses than a full merger. What is rare – what has the odds stacked against it – is a full merger, with the unification of the two gene sets, and the extra genes being used for complexity rather than being disposed of or coopted for conflict within the cell. The complexity of life has evolved more by piecemeal duplications of genes within DNA molecules than by grand mergers. But one grand merger did succeed, and it was a turning point in evolution. All subsequent complex life has been shaped by the eukaryotic revolution.

I have compared the eukaryotic cell to a business merger, but the subtle force of selection in the shrinking mitochondria does give me pause. People who work in business are understandably reluctant to give away their jobs or surrender control. Perhaps we should think instead of some sphere of life in which self-sacrifice is more reasonable. Perhaps the early eukaryotic cell was more like a monastic scriptorium shared by the scribes of two religions. When one scribe leaves the room, who knows what may happen while he is out. Maybe his copybook catches fire. Maybe his inkpot spills over the morning's work, or his quill snaps in two, or his tame cockerel sickens and dies the following night with a sad eye and ruffled feathers. The whole set-up is unsustainable. It will probably come to an end when one scribe is martyred or enslaved. But a subtle scribe may realize that the true goal of religion lies in the propagation of its doctrine, not in the manual work of the copyist. He will not do anything crude, messy or savage when he is left alone in the scriptorium. He will instead, with pious irony, discreetly correct the heretical master-copy and its sacrilegious illustrations. He will progressively turn the other scribe around, until that scribe has become the ghostwriter of the true religion. When, after many years, he has successfully incorporated his religion into the formerly competing manuscripts, his temporal work is complete and he can seek a martyr's crown with good cheer. He may go to the stake in a manner that onlookers will find quite edifying. Two thousand million years of evolution in the eukaryotic cell have resulted in a dominant nuclear message and a tiddly organelle message which is a shadow of its former self. But we should beware of condescending to this organelle message. Even the shadow retains the power to surprise, and self-castrated hermaphrodite flowers, declining muscle-power in ageing humans and disintegrating chloroplasts in one-celled isogamous algae all bear witness to the crucible in which the synthetic, complexity-boosting, eukaryotic cell evolved after its long Darwinian delay.

7
The justice of the peas

The byter bitten

Sexual reproduction is a great source of conflict. It means that two unrelated sets of genes are forced to cohabit in one offspring body. They not only have to cohabit, they have to cooperate, at least in complex forms of life. But it is not at all obvious that they will. There are all too many opportunities for selfishness. Some of the genes may evolve to exploit the body for their own advantage rather than cooperate with the other genes to produce a superior body. The opportunities will be seized if at all possible because the advantages can be enormous. In Chapter 6 we thought about how natural selection acts on the genes inside a new sexually reproduced cell. We concentrated there on the genes in the mitochondria and chloroplasts. Now we can think about the same question, but for the nuclear genes. The new cell contains two versions of every gene, one from each parent. The two will be passed on equally to the future offspring if they allow themselves to be passively copied. But a selfish gene may be able to give itself an unequal – and doubled – chance of being passed on.

Imagine a gene. It codes for an enzyme, perhaps one of those DNA repair enzymes that we met in Chapter 4. The DNA repair proceeds by several steps; one of them is to digest any erroneous bases, and the gene in question might perform this task. The gene is positioned at a certain site in the DNA, and the cell contains two copies of the gene, at equivalent positions in the two DNA molecules that it inherited from its two parents. Mutations in one of the copies of the gene – the maternal copy, perhaps – might change it from coding for a tame repair enzyme to a destructive one, giving it the ability to 'repair' (in an ironic sense) the equivalent gene in the paternal DNA in the same cell: it bites out the other gene and replaces it with a second copy of itself. This would be a neat piece of genetic selfishness. Before the 'repair', the maternal gene had only a 50 per cent chance of being passed on in future reproduction. Afterwards it has a 100 per cent chance. Natural selection automatically favours it. The tame and destructive versions of the gene are in evolutionary conflict, because there is counter-selection on the paternal gene not to be bitten, and even to bite back.

The repair enzyme example is merely expository and is probably too good to be true. A real gene, as we shall see, probably could not evolve to copy itself precisely into the other version of itself in the same cell. But there are cruder possibilities that are eminently plausible. A gene on one chromosome could evolve to accelerate that chromosome's reproduction, or to destroy the other copy of the same chromosome; either way, the chromosome would take over the cell, as we saw in Chapter 6 (Figure 8). The same problems that we noticed there also apply to disruptive nuclear genes. One problem is that the two copies of a gene in a sexually reproduced offspring cell are very similar. One copy is redundant and there may be little direct, immediate disadvantage to a gene that destroys and replaces the other copy of itself. Even if one gene set eliminates the whole of the other gene set, it is not as bad as might naively be thought – it is not as bad as the left half of the body destroying the right half of the body. But a destructive gene of this sort nonetheless endangers the life form that it is in, both because it does incidental damage and, over the longer term, because two destructive copies of a gene might be inherited by the same cell. They could mutually destruct, and kill the cell. Moreover, a life form that allows destructively selfish genes to spread inside itself would be unlikely to evolve to be complex. A DNA repair enzyme, for instance, could theoretically be used in the evolution of complexity. A more complex life form might be one that lives longer, and one way it could live longer would be by repairing its DNA more carefully. However, the advantage of living a little longer would probably be small, and easily outweighed by the double advantage of the other mutant version – the destructively selfish version – of the DNA repair enzyme. The complex life forms that we see on Earth today will have evolved from ancestors in which selfish internal subversion was somehow prevented.

In Chapter 6 we saw one mechanism that tames genetic selfishness. Natural selection is impossible between identical genes, and natural selection within a body can therefore be prevented by imposing genetic uniformity. The genes will then evolve to cooperate. This method works for organelles such as mitochondria. The mitochondria of a cell cooperate because they reproduce clonally, and only one parent's mitochondria make it into the offspring. All the mitochondria in the body, at least initially, are the same. But the same method cannot be used for the nuclear genes of complex life. The DNA of a mitochondrion is so small that a copy of it usually does not contain a mistake, and natural selection can remove the small number of mistaken versions. Sex is unnecessary. In the nuclear DNA of a complex life form, mutations are far more numerous and the nuclear DNA would suffer a mutational meltdown if it was reproduced non-sexually. The DNA of an individual can be saved, as we saw in Chapter 5, by importing another individual's DNA. The two individuals' DNA will contain different mutational errors, and complementary good and bad parts; they can be combined to create a new DNA molecule with fewer errors than either parent. Genetic diversity is unavoidably needed for this process. It would not help to import a second DNA molecule that was identical to the one already present, or to make the two identical: the two would then contain the same errors. It would be like (to allude back to Maynard Smith's analogy in Chapter 5) swapping the ignition systems between two vehicles that both contained defective ignition systems. Sex helps only if you swap parts between two vehicles that have different defects. Indeed, almost any explanation for sex will require that the two parental DNA molecules differ. Kondrashov's theory may turn out wrong, and some other theory will be needed to explain sex. This theory, whatever it is, will almost certainly not work for the imaginary case in which the father's and the mother's DNA is identical. Sex is now pointless because cloning can achieve the same results. Cooperation has to be achieved between the maternal and paternal genes in the nucleus, but it cannot be by forcing them to be identical. Chapter 6 described one route to genetic cooperation, but it cannot be the only one.

In this chapter we shall look at another route. The principle is that the genes cooperate because the information that they would need to act selfishly has been randomized away. Imagine two armies, one made up of soldiers who wear red uniforms and another made up of soldiers who wear black uniforms. Sex generates an offspring that is potentially a battlefield for these two armies. Evolution has found two ways to stop the fight. One is to make everyone wear the same colour of uniform: this was the method of Chapter 6. The other is to randomize the uniforms between the two armies. You make half of one army change their uniforms from red to black, and half the other army change from black to red. A soldier is now as likely to shoot a comrade-in-arms as an enemy; he does not know which is which. This chapter will be concerned with how the genes have evolved in a way that is analogous to this uniform-swapping.

War and peas

In 1805, the land a few miles outside Brünn (now Brno, in the Czech Republic) on the way to Austerlitz (now Slavkov) was the theatre for one of Napoleon's biggest triumphs, in which he smashed the alliance of the Austrian and Russian emperors. The battle of Austerlitz is well known to historians and to readers of Tolstoy, but I doubt whether our lives would be all that different if the battle had never been fought. The real claim of Brno to a place in history lies in events some fifty years later, when Gregor Mendel began to grow peas in the monastic garden there. Those peas flowered at one of the great moments, or turning points, of history: they revealed to Mendel the principles of heredity and led to genetics, to Watson and Crick, the human genome project, and all modern and future gene technology. Until recently Mendel had inspired no novels, and was abandoned by historians to provincial obscurity; but the tide of interest is starting to turn. Mendel and his assistants counted tens of thousands of peas, and I expect that, in fifty years or so, educated people will agree that Napoleon was historically no more significant than any one of them.

Mendel's actual research combined brilliant experimentation with brilliant reasoning; you will not find many other biologists using the theory of probabilities in the nineteenth century. We now know about Mendel's work in terms of two laws – the laws of segregation and of independent assortment – that do summarize part of his work but were not invented until after his death. The two laws provide the link between Mendel and modern genetics, and we now understand them in terms of genes. I am going to look at the law of segregation first. (I shall also refer to the law of segregation as the first law, and to the law of independent assortment as the second law.) Mendel's insight was that the attributes of an individual are controlled by pairs of hereditary factors (that is, genes), inherited from the two parents. The genes are preserved, unaltered, during the individual's lifetime. When the individual comes to reproduce, each pair of genes divides (or 'segregates') and one gene from each pair goes into any one gamete.

Flower colour was one of the things that Mendel studied; he concentrated on the purple and white blossoms of the pea plant. These colours, in peas, are determined by two versions of a gene, and Mendel introduced the convention that is still followed in which we symbolize the different genes by different letters, such as P and p. A pea plant that inherits two P genes (symbolized by PP) or one P and one p gene (Pp) has purple flowers; a pea plant that inherits two p genes (pp) has white flowers. The law of segregation tells us that half the gametes produced by a Pp plant will contain P and half will contain p. The law is probabilistic rather than exact. If a Pp plant produces ten offspring, the law tells us that on average five will contain P and five p, but in any one ten-offspring family a gene may be in four or six, or even three or seven, offspring. It is as if a coin is tossed to decide whether an egg will be fertilized by a pollen with one gene or the other: 'heads' for P, and 'tails' for p.

The law of segregation underlies complex life. It tells us that the inheritance of genes is in fact just and equal: each gene of a pair has the same 50 per cent chance of being passed on. Complex life really is built on the fair inheritance of genes. But this still leaves us with the question of why it is fair. Laws were made to be broken. The law of segregation is not something that just happens and needs no special explanation. Some positive mechanism will have been needed to enforce the Mendelian discipline. What is the mechanism? One way to find out how human laws are enforced is to watch what happens when people break them. We can look now at some genes that seem to break Mendel's law, and see what they tell us about Mendelian law enforcement.

Mendelian lawbreakers

Several genes are known to break Mendel's first law. The classic, and best understood example is a gene in fruit flies called 'segregation distorter'. It was discovered in the 1950s and has since then been found in many (but not all) populations of fruit flies. Even where it is found, it is rare, and most fruit flies continue to obey Mendel's law. I am going to refer to the segregation distorter gene as an 'assassin' gene. A normal gene has a half chance of being passed on to an offspring; the assassin gene has more than a half chance because it kills the other, normal, version of the gene. I shall call the normal version of the assassin gene a 'victim' gene. Most fruit flies have two victim versions of this gene – one from the father, the other from the mother – and their inheritance is perfectly normal; indeed it is hardly helpful to call them 'victim' genes in their normal set-up. But some fruit flies inherit an assassin gene from one parent and a victim gene from the other; then the name makes sense. Mendel's law states that half the offspring of these flies should inherit the assassin gene and half the victim gene. In fact almost all the offspring inherit the assassin gene. The particular example of the segregation distorter gene works in males only. The gene pulls its trick during sperm manufacture. Half the sperm (the half containing the victim gene) have an accident. The chromosomes in the sperm fail a crucial folding routine and the sperm do not develop. The result is that almost all the male's sperm, and his offspring, contain the assassin gene.

Does natural selection favour the assassin gene? The assassin gene clearly does better than the victim gene in male fruit flies that contain both versions of the gene. But we also need to consider its success relative to the population as a whole. The assassin gene kills half the sperm, and the sperm count of a fly that contains an assassin gene is half that of a normal fly. A halved sperm count does not mean halved reproductive success, however, because a male fly's reproduction is limited more by the number of female flies he can jump on than by his sperm supply. The reproduction of a male with an assassin gene is reduced, but to maybe 60 per cent of the normal level, not 50 per cent. Natural selection therefore favours the assassin gene. Suppose that a normal male produces a hundred offspring (if we count offspring at the fertilized egg stage); fifty of the offspring will contain one of the copies of each of his genes and fifty will contain the other copy. The fair Mendelian share for the assassin gene would be fifty offspring. But the assassin gene reduces the male fly's total reproductive output to sixty, of which all sixty contain the assassin gene. The number of offspring with the assassin gene goes up from fifty to sixty: the assassin gene gains by taking a larger piece of a smaller pie.

Geneticists have studied the segregation distorter gene of fruit flies in some detail, and found that what was originally thought to be one gene is in fact a combination of two main genes, lying side by side on the chromosome. Fruit flies have a unit set of four chromosomes, like our set of twenty-three chromosomes. Theirs are named chromosomes 1 to 4, and the distorter gene is on chromosome 2. One of the two genes is the assassin/victim gene. The other is a gene used for recognition; it also has two main versions, which I shall call the 'target' and 'safe-conduct' versions of the gene. Most fruit flies inherit the target version of the gene from both parents, and again the term is not helpful for a normal fly. But we are concentrating here on the action of the rare assassin genes, which tend to be associated with the safe-conduct genes. The assassin gene does its damage in a male fruit fly that inherits a particular pair of chromosomes. It must inherit the assassin gene and the safe-conduct gene on the copy of chromosome 2 from one of its parents and the victim gene and the target gene on the copy of chromosome 2 from the other parent. The assassin gene then kills the chromosome that contains the victim gene by aiming at the target; it leaves the chromosome with the safe-conduct gene in peace. The assassin does not kill its victim directly: it aims at a target gene on the same chromosome as the victim, and destroys the whole chromosome.

Why does the assassin gene not kill the victim gene directly? The second gene (the target/safe-conduct gene) is probably needed because of a problem of recognition. A cell nucleus in a fruit fly contains eight chromosomes, four from each parent. The assassin gene is sitting on one of the copies of chromosome 2. It works by killing whole sperm, and natural selection favours it only if it selectively destroys sperm that contain the other copy of chromosome 2. The assassin gene has to recognize the copy of chromosome 2 that it is not on. If it aims at either of the copies of chromosomes 1, 3 or 4 it will kill sperm that contain itself 50 per cent of the time; if it aims at the copy of chromosome 2 that it itself is on, it commits suicide 100 per cent of the time. How can the assassin gene recognize the one out of the eight chromosomes that it is after? Genes do not act directly, but code for other molecules that carry out the action; DNA has the intermediary molecule called RNA transcribed from itself, and the RNA is used to assemble a protein. The protein then does something, such as recognizing a target. A simple assassin gene could produce a protein that binds to the other copy of chromosome 2 and then causes it to be destroyed. Somehow the molecule produced by the assassin gene must float around the nucleus, recognize the target gene and give it the molecular Judas kiss. The recognition problem arises because the molecule produced by the assassin gene is floating in a soup of chromosomes and other molecules that influence the chromosomes. The chromosomes are all rather similar. They are all made of DNA, together with proteins that are bound to the DNA. It is not at all obvious that a gene-binding, would-be Judas of a gene product will selectively bind to the right chromosome out of the eight available.

The recognition problem is odd from our perspective. We have no problem in distinguishing ourselves from others. If I give you a gun and tell you to shoot people other than yourself, I expect you to get it right. You can start by shooting me. I do not expect you to dither, undecided whether to blow your own head off or mine. This is a case in which it can be misleading to imagine a gene as a personal agent. We might overlook the crucial problem of recognition. A person can effortlessly recognize itself, as distinct from other people. But things are not so easy for a gene that has to distinguish the chromosome it is on from the other copy of the chromosome. A gene does not have mental skills. It is just a sequence of mindless nucleotides, with the power, at most, to code for a protein. The protein might have the ability to destroy DNA, but it is like a loose cannon on deck if it cannot recognize which DNA to destroy.

The solution is to find some sequence of DNA unique to the other copy of chromosome 2 – this sequence is the target gene. The target sequence must have two versions that differ between the two copies of chromosome 2. The assassin gene can then evolve to bind the other copy of the chromosome, not its own copy, and the recognition problem is then solved. Target sequences of the right sort will not be easy to find, however. The two chromosomes of a pair are necessarily similar. They have the same set of genes and differ in little bits of sequence only. The target has to be found in one of these little bits. We know something about the 'target' sequence of the fruit fly's assassin gene. It is a stretch of DNA made up of a number of side-by-side repeats of a unit sequence; the unit sequence is 120 letters long. The versions of the target that are successfully attacked by the assassin gene contain hundreds or even thousands of repeats of the unit sequence. The safe-conduct version contain few, or no, repeats. Repeat sequences are scattered about the DNA, for one reason or another, and may be natural targets of assassin gene action. A sequence that is repeated in the DNA will be easier to find. If you have a pile of holy manuscripts and your job is to find the heretical copies and burn them, your task will be easier if the heresy is repeated than if it appears once and makes only an iota of textual difference.

What is the targeted repeat sequence there for? It presumably does something useful when it is not getting its chromosome killed – but we do not know what. We do have evidence that it is useful. An experimental population of fruit flies can be set up in which none of the flies has the assassin gene; some of the flies have the safe-conduct version of the repeat sequence and other flies have (like normal natural fruit flies) the target version with many repeats. If the repeat sequence did not matter, the frequency of the safe-conduct and target versions should wander about at random over time. In fact the target version, with high repeat numbers, zooms up in frequency, indicating that it is advantageous for a fruit fly to have multiple repeats of the 120-letter unit in the absence of the assassin gene. The evolutionary spread of assassin genes does damage not only by partly castrating the fruit flies but also by setting up a selection pressure in favour of genes that would otherwise be inferior.

We can now see more clearly why complex life could not survive without Mendelian gene justice. One assassin gene is damaging, but tolerable. A male fruit fly containing the assassin gene has his fertility reduced, but he still breeds and the population survives. Fruit flies continue to exist. But a fruit fly contains 14,000 or so genes: imagine what would happen if a few more cracked the system in the same way. One such gene killed half the fly's sperm. A second, provided it was on another chromosome, would kill half the remaining sperm, and a fly with two independently active assassin genes would have 25 per cent of the normal sperm supply. You would not need many of the 14,000 genes to turn nasty in this way for the fly to be genetically autocastrated. No life form could exist in which many genes had found a way round Mendel's law. (Mendelian distortion is equally advantageous in females. Plenty of assassin-like genes operate in females, but they are not as well understood as the assassin gene that screws up sperm manufacture and troubles only males.) The problem probably grew worse as more complex life forms evolved. More complex life means more genes, and more genes means more potential sources of damaging assassin molecules and more potential target sequences for the assassin genes to aim at. Assassin-like genes have probably been a recurrent problem during evolution, and a problem that potentially increased as life became more complex. And yet most genes, in us as well as in fruit flies, do obey Mendel's law. The question is, why?

The answer almost certainly lies in the damage that the assassin gene inflicts on the other genes in the body. The assassin gene, we saw, partially autocastrates the fly. It shrinks the size of the total reproductive pie. Natural selection favours the assassin gene itself because it is carried by all the sperm rather than half of them. But what about the other genes? The victim gene of the assassin–victim gene pair is the big loser: it gets into none of the sperm. Any other gene that is near the victim gene on its copy of chromosome 2 experiences a similar fate. Genes on the other three chromosomes do not do as badly as the genes on the victim gene's chromosome, but they still lose. They get into the standard Mendelian fraction (one half) of the sperm, and their reproduction is reduced because the total sperm supply is reduced. Natural selection works strongly on the victim gene to suppress the assassin gene, but it is only one gene and may not evolve fast enough to save itself. It is the other genes that are potentially a more powerful force to suppress the assassin gene, because there are so many of them. Natural selection on most genes other than the assassin gene will favour any ability to suppress the assassin and restore full fertility.

In fact an amazing array of genes are known to influence the assassin gene. Some genes can suppress, or partly suppress, it; others can enhance its effect, or suppress the suppressors. The genes are predictably positioned: the suppressors tend to be some distance away from the assassin gene in the DNA, and the counter-suppressors tend to be near to the assassin gene. It is not known how these individual genes work, but they could produce a specialized molecule that binds to the assassin molecule and neutralizes it, or one that sits on the target sequence and protects it. These molecular games probably go on, but I doubt whether they explain Mendel's first law. The weight of numbers should give suppression the upper hand in the evolutionary conflict, but inheritance at any moment will depend on the messy details of which genes have evolved which particular abilities. The assassin gene will work by one detailed molecular mechanism and will require one kind of suppressor; another, similar gene will require another kind of suppressor, and so on. The net result might be fair segregation at some times, but it is not obvious that it would produce the kind of steady, predictable Mendelian justice that we observe. It is also not obvious that complex life could exist if large chunks of the DNA were given over to subversion and counter-subversion. Life would become like East Germany before the fall of the Berlin Wall, with one police informer per 6.5 citizens. We should therefore be on the look-out for some more general mechanism of Mendelian law enforcement. It should consistently produce the fair inheritance of genes, not require lots of messy and detailed gene interactions, and not implicitly coopt most of the DNA to protect, rather than code for, complex life. One such mechanism has been proposed recently by James Crow, and by David Haig and Alan Grafen.

The second law

Mendel's first law – the law of segregation – applies to any one gene pair, controlling one feature such as flower colour. His second law – the law of independent assortment – applies when we consider more than one pair of genes. For instance, Mendel studied the inheritance of height as well as flower colour. Some pea plants are tall (about 1 metre), others dwarf (about 0.5 metre). Height, like colour, is controlled by one gene with two versions: TT and Tt plants are tall; tt plants are dwarf. Now consider a tall purple pea plant, with the gene set $TtPp$. The law of segregation tells us that for the height genes half the gametes will have T and half t, and for the flower-colour genes half the gametes will have P and half p. But how will the two be combined? Maybe half the gametes will be TP and half tp. Maybe half will be tP and half Tp. The law of segregation would be satisfied either way, with respect to both the flower-colour genes and the height genes. In fact the two gene pairs are inherited independently. A quarter of the gametes contain TP, a quarter Tp, a quarter tP, and a quarter tp. When a gamete is formed, the Mendelian coin is first tossed to decide whether to put a T or a t in it and is then tossed again to decide about P and p. Two heads in a row, and the gamete is TP; a head and then a tail, and it is Tp; a tail and then a head ... and so on. The law of independent assortment says that the genes controlling

177

different features are inherited independently in the offspring. The flower-colour and height genes are not inherited as a unit.

The genes are inherited independently because of the process of genetic recombination, a process we have met before. Recombination is the mechanism that shuffles genes during sexual reproduction. It is the genetic mechanism on which the sexual motor mechanics analogy used in Chapter 5 was based: I wrote about shuffling good and bad components, such as ignition and brake systems, between cars. In life, gene sets are shuffled by genetic recombination. Recombination happens for two reasons. One is that the two gene pairs are carried on different chromosomes. Peas have seven pairs of chromosomes, numbered 1 to 7; the genes controlling flower colour are on chromosome 1 and the genes controlling plant height are on chromosome 4. A pea plant might inherit a P and a t gene on the chromosomes 1 and 4 it inherits from its mother, and a p and a T gene on the chromosomes 1 and 4 from its father; but when it reproduces, each of the seven originally maternal chromosomes are passed on independently. The originally maternal chromosome 1 (with P) is equally likely to enter a gamete with the originally paternal chromosome 4 (with T) as it is to enter a gamete with the originally maternal chromosome 4 (with t). This shuffles the genes on different chromosomes. The plant received a TP combination from one of its parents, but this T is equally likely to be passed on with the p from the other parent as the P it came in with.

The second reason why recombination happens is that the maternal and paternal copies of each chromosome physically swap some of their genes. The DNA chains of both chromosomes break at the equivalent point, and the paternal end of one joins to the maternal end of the other. This shuffles the genes on either side of the break point. Genes at opposite ends of the two chromosomes are frequently shuffled, because a break nearly always takes place between them. But genes close together are rarely shuffled and more often inherited as a unit. Genes close together (or 'linked') on a chromosome are inherited as a unit most of the time: if Mendel had studied pairs of closely linked genes he would not have discovered independent assortment – or he would have discovered a more complicated law in which some genes are inherited independently and some are not.

Crow, and Haig and Grafen, argued that recombination acts to suppress assassin genes in general. We can see the argument in the particular example of the fruit fly assassin gene. We have thought so far about two kinds of chromosome, with respect to this gene complex:

(1) *assassin gene* and *safe-conduct gene*
(2) *victim gene* and *target gene*

The system works (from the assassin gene's viewpoint) provided these two are the only gene combinations; then the assassin gene can kill sperm that contain victim genes by detecting the target gene. But these two are not the only kinds of chromosome that are logically possible. The other two possibilities are:

(3) *assassin gene* and *target gene*
(4) *victim gene* and *safe-conduct gene*

Possibility (3) is particularly unfortunate for the assassin gene. The chromosome is suicidal; the assassin gene attacks itself. The last combination also, if less drastically, neutralizes the advantage of the assassin gene; it makes the victim gene invulnerable. If these two chromosomes spread among the fruit flies, natural selection will work against the assassin gene. The two genes have to conspire together in the right combination (the assassin with the safe-conduct gene) if the course of Mendelian justice is to be satisfactorily perverted.

A population might initially have only the first two kinds of chromosome. Natural selection favours the assassin gene, and Mendelian justice starts to collapse. But it can be saved by genetic recombination. Recombination shuffles the genes and the assassin gene becomes combined with the target that it attacks. Natural selection now works against the assassin gene and Mendel's law is reimposed. We can therefore think of a gene that causes recombination between the assassin gene and its target as a suppressor, in an evolutionary sense, of assassin genes.

Recombination randomizes information in the DNA, information that might be used to distinguish the two chromosomes of a pair. The target sequence initially was like a big banner saying NO COPY OF ASSASSIN GENE ON THIS CHROMO-SOME. The assassin gene acted accordingly. Recombination removes this information, creating uncertainty about whether a target sequence is on the same chromosome as the assassin gene, or a different one. It is a wonderfully effective general Mendelian law enforcer. I said at the beginning of the chapter that we were going to look at mechanisms that resembled the swapping of uniforms between two armies. We now see that the mechanism is even more cunning than that. After the gene shuffling, the assassin gene is not just ignorant about who is friend and who is foe; it is also ignorant about who it is itself. The assassin gene has been forced to play Russian roulette with 50 : 50 odds.

Gene shuffling will be a general anti-assassin mechanism if assassin genes generally work by aiming at an associated target gene. Arguably they will. The reason is the recognition problem. If a gene could recognize directly the other copy of itself, bind to it and destroy it, shuffling would do nothing to stop it. There is nothing logically impossible about such a gene and, who knows, they may crop up from time to time and whiz through the population unstoppably. But they are biologically unlikely. Such a gene is only a small part of the chromosome, and will hardly differ at all in its sequence from the gene that it is trying to kill. Indeed, the assassin gene would have to recognize a stretch of the DNA that was more like itself than any other sequence in all the DNA; the sequences of the two genes might differ only in one nucleotide. There would be a danger of recognition error – and error means suicide. Moreover, potential recognition sequences are much more numerous outside the gene than in it. The recognition problem is so big that shuffling probably works against nearly all lawbreakers. And if this seems too extreme a statement, then at least we can say that recombination protects against a large class of potential lawbreakers. All known lawbreakers that have been analysed in sufficient detail do contain more than one gene. (We shall look later at some genes that are analogous to, but not conventionally recognized as, segregation distorters; they consist of only one gene but can work only in special circumstances. We shall also meet some single-gene lawbreakers in the next chapter, but they can only be mildly damaging.)

The theory makes sense of many detailed facts about the assassin gene complex. One is the distribution of suppressors, and enhancers, of the assassin gene. I mentioned earlier that other genes that suppress the assassin gene tend to be located well away from it in the DNA, whereas other genes that enhance the assassin gene (or suppress the suppressors) tend to be near to it. This makes sense. If you are a gene sitting right next to the assassin gene, recombination will practically never split you from it; you will be inherited with it as a unit and gain the same advantage as the assassin gene does from eliminating the sperm that contain the other copy of the chromosome. But if you are a gene miles away elsewhere in the DNA, you have – because of recombinatorial gene shuffling – a 50 per cent chance of being in the sperm with the victim gene and a 50 per cent chance of being with the assassin gene: you are the kind of gene that on average loses out. The assassin gene evolutionarily corrupts the genes in its immediate vicinity, but inspires more distant genes to acts of positive Mendelian virtue. A second good fact is that the assassin gene and target genes in the fruit fly are in a peculiar chromosomal region – peculiar in that recombination is locally impossible. The genes are located in an inverted region: a region where the DNA is spelled backwards relative to a normal copy of chromosome 2. Recombination usually fails inside inverted regions and the assassin gene therefore cannot be split from its recognition sequence. The assassin gene has probably been able to spread only because it happens to be sitting in an inverted stretch of DNA. In general, we can expect groupuscles of assassins to hide themselves in the bits of DNA where the genes are not shuffled and whole sets of genes are inherited as a unit. Mendel's second law, the law of gene shuffling, evolutionarily acts to enforce the first law, the law of fair heredity, and where gene shuffling is prevented gene justice breaks down.

A set of Mendelian lawbreaking genes, such as the assassin gene and the safe-conduct gene, are like a conspiracy against the state in human politics. Machiavelli supplied a superbly 'Machiavellian' analysis of conspiracies in the *Discorsi*. Here is the classic historian of the Italian Renaissance, Jacob Burckhardt, for the background: 'Machiavelli, in his famous *Discorsi*, treats of the conspiracies of ancient and modern times from the days of the Greek tyrants downward, and classifies them with cold-blooded indifference according to their various plans and results.' Machiavelli makes various points about the successes and failures of conspiracies, but his most powerful point (and the one that is most relevant here) concerns the number of conspirators. The assassin's problem is that the prince will be protected by guards. A prince is always in danger from lone assassins, who are so desperate that they are willing to die in the attempt. 'A poor, miserable Spaniard stuck a dagger in the neck of Ferdinand, King of Spain, and, though the wound was not fatal, it shows that a man of this type may have both the intention and the opportunity of doing such a thing.' The poor Spaniard, acting alone, knew that he would die, but most assassins will prefer to survive, and this makes it necessary to involve more people in a conspiracy: one or more to do the killing, one or more to secure the get-away. In human politics, the need for multiple conspirators is to overcome the prince's entourage and provide defence against later reprisals. It is not a need for recognition: a single assassin can recognize his princely victim directly. But Mendelian and political subversion share an abstract similarity in the need for multiple conspirators, even though the exact reason for it differs. Moreover, in the full 'segregation distorter' system of fruit flies there are more than two genes, and the action of some of the other genes in suppression and counter-suppression is probably more analogous to the sword-clash of conspirators and guards in princely politics.

A conspiracy will potentially be more deadly as it includes more people, but larger conspiracies almost always fail. The reason is that they are found out before they go into action. Informers reveal the plan, either from 'lack of loyalty' or 'lack of discretion'. A conspirator may calculate that he can gain by grassing: when you try to talk some malcontent into joining your conspiracy, 'by the very fact of your having opened your mind to such a malcontent, you provide him with the material with which to obtain contentment.' The prince will reward that malcontent, when he or she sneaks on you. Or a conspirator may talk carelessly 'to a lady friend or to a boy friend or to some other frivolous person, as did Dymnas, who with Philotas and others conspired against Alexander the Great, and talked of the conspiracy to Nicomachus, a boy of whom he was fond, who at once told his brother Cebalinus about it, and Cebalinus told it to the king.' Frivolous lovers and calculating malcontents find their analogy in the gene shuffling of Mendel's second law. An assassin gene cannot trust a safe-conduct gene to stay with it for long. Likewise, individual human beings move around and interact with many other people, including friends of the state. The ordinary fluidity of social networks shuffles up individuals and prevents conspiracy in much the same way that genetic shuffling prevents subversive conspiracies in the Mendelian body politic. Machiavelli found that the best hope for a conspiracy is if it stays small, binds itself as a unit away from the outside world, and keeps the planning stage as short as possible. Princes, in turn, can make conspiracies less likely by discouraging small assemblies among their subjects. Gene recombination is more effective than its social equivalent, and the organization of a body, and cooperation between its parts, is much greater than in a society. Machiavelli himself would be pleased. He relished social conflict, as a sign of vitality. He would have been alarmed by a mechanism that successfully reduced us to social uniformity. Others will disagree, but we can all be grateful that our ancestors 2000 million years ago evolved a mechanism that effectively suppresses genetic conspiracies within our DNA. Without it, we should not exist.

Mendel's second law prevents genes from acting in concerted multigene units. In some cases this may be a disadvantage, because one gene might be able to form a specialist relationship with another in a way that was beneficial for the body. Complex life has had to deny itself these benefits. The DNA of bacteria does seem to contain linked sets of more than one gene that work together as a unit, and are (at least sometimes) beneficial to the bacteria they occupy. But in complex, sexually reproducing life forms, a multigene concert is all too easily corrupted into selfish conspiracy. Complex life has a mechanism, in the gene shuffling procedure, that prevents one gene from associating with another over evolutionary time. An individual gene is passed on as a unit during evolution. Within a body, one gene cooperates with the other genes to build, defend and reproduce the body, and the cooperation between genes is far higher than any cooperation we see between individuals in social life. But the cause of this genic cooperation is the atomistic individualism and independent inheritance of the genes as they are passed from one body to the next. If the genes were not forced by Mendelian inheritance to be units, they could become conspirators, and complex life with its thousands of cooperating genes would be a science fiction.

Russian roulette in the sorority

Meiosis – the special cell division that produces sperms or eggs – halves the number of genes in each cell. We are now familiar with the problem that the halving creates: natural selection acts to favour genes that increase their chances of being passed on to more than their fair Mendelian share of 50 per cent. Gene shuffling worked against multigene assassin systems, but they are not the only perverters of Mendelian justice. We can introduce the next kind by looking at a puzzling detail in the way that gene numbers are halved during meiosis. Our cells normally contain two sets of chromosomes. The total number of chromosomes in a cell can be written as $2n$; n is the number of chromosomes in a unit set, in a sperm or an egg. In humans n is 23 chromosomes and $2n$ is 46 chromosomes; in fruit flies, n is 4 chromosomes and $2n$ is 8; in a pea plant, n is 7 chromosomes and $2n$ is 14. If the purpose of meiosis is to halve gene numbers, we should expect it to do just that: to halve them. Meiosis ought to proceed as follows:

$$2n \rightarrow n$$

This would be called a one-step meiosis, and oddly enough it is not what happens. Instead meiosis goes like this:

$$2n \rightarrow 4n \rightarrow 2n \rightarrow n$$

Meiosis is meant to halve gene numbers. So what does it do? It starts, of course, by doubling gene numbers.

Meiosis has two steps following the apparently unnecessary round of gene doubling. Here I want to consider a possible explanation, put forward by Haig and Grafen. They suggest that the extra step prevents the spread of a theoretical, destructive kind of gene that they call a 'sister-killer'. The language here is as follows. When one cell divides into two, we call the two cells 'daughters' of the 'mother' cell, or 'sisters' of each other. A sister-killer gene in one cell destroys its sister cell after cell division (Figure 9). Natural selection will favour a sister-killer gene if it is in one of the cells but not the other. The trouble with a one-step meiosis is that it creates exactly that condition. We first suppose that some gene with the sister-killer ability arises by mutation: it can recognize, or mark, the sister cell and either destroy it or cause it to

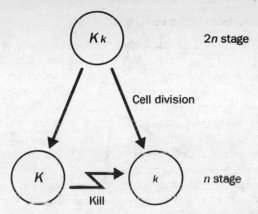

Figure 9. One-step meiosis is vulnerable to sister-killer genes. The cell initially has two versions of each gene, indicated by *Kk*; *K* is a sister-killer gene. The two genes are reduced to one in the cell division that produces the gametes; the sister-killer gene will be in only one of the two gametic cells after cell division. It 'knows' it is definitely not in the other cell and natural selection favours it when it kills the other cell during or immediately after cell division.

be destroyed. This is not all that remarkable an ability. The sister-killer gene might kill directly, or indirectly by coopting one of the many bodily mechanisms for destroying unwelcome cells. These mechanisms recognize cells that need to be destroyed by various means; for example, they might detect that the cells are leaking vital chemicals. Normally it would only be a wounded cell that was leaky, but a sister-killer gene could squirt the tell-tale chemicals on its sister cell. The sister-killer gene, however it works, will be a rare new gene when it first appears. Most members of the population will not contain it, and some will have a single copy of it together with a copy of the gene that it victimizes. Then, when the cell divides and the genes are halved, the sister-killer gene is in only one of the two cells. It increases its frequency among the gametes when it kills the other cell. It gains a larger share of the reproductive pie. The action of the sister-killer genes may shrink the total size of the pie, but natural selection still favours them because they are taking more of it.

The sister-killer gene is a single gene that breaks Mendel's law

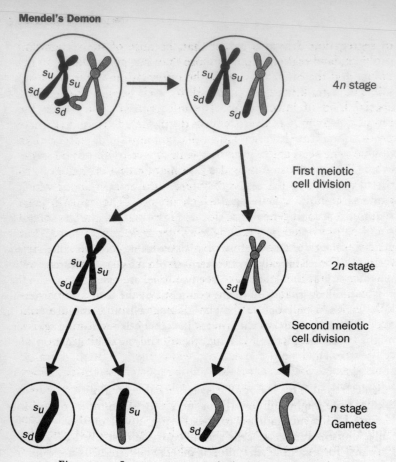

Figure 10. In a two-step meiosis, a gene is made uncertain whether a copy of itself is in the same cell as itself, or in a sister cell. We start with the cell at the top left. Meiosis begins by doubling up all the genes; this has already happened by the time of the picture. There are two (attached) copies of the maternal chromosome (black) and two (attached) copies of the paternal chromosome (grey). A paternal and a maternal chromosome may now undergo recombination, such as at the site where the black and the grey chromosomes have come together. In the example here, the sister-killer gene has been inherited on the maternal chromosome. We need to compare the fate of a sister-killer gene that happens to be upstream (s_u) or downstream (s_d) from the site of recombination. The upstream s_u gene is in one

of segregation. I argued earlier that, because of the recognition problem, lawbreaking requires more than one gene: one to do the killing and the other to provide the information about whom to kill. The sister-killer gene can work by itself because it exploits a special kind of information. The information comes from the physical proximity of the sister cells during cell division. In gamete production, the sister gametic cells contain single rather than double gene sets, and a gene that is in one cell will not be in the other. The cell division provides information that is analogous to the target gene in the assassin system: information about which gametes contain a gene and which do not. The assassin gene required a second gene to provide this information, and its spread could be prevented by shuffling the killer gene and its target. But gene shuffling does not help with a sister-killer gene in a one-step meiosis. The sister-killer gene gains from killing the other cell whether or not the genes have been shuffled.

It is unlikely that a life form could survive if it allowed sister-killer genes to evolve. Exactly what damage would be done would depend on the details of the sister-killer. The first sister-killer genes might be inefficient, and manage to kill only a small fraction of cells; life would be not much the worse for it. But the damage could then escalate, as natural selection replaced the initial sissified killers with more powerful versions. The most virulent sister-killer gene could kill half the gametes, and create the kind of snarl-up that we met before. One such gene is bad enough, but evolution could then conjure up another, on another chromosome. Once the population contains a few independently operating sister-killer

of the two cells after the first cell division, but in both the cells produced by the second division. Natural selection only favours s_u genes if they kill after the first cell division. The downstream s_d gene is in both cells after the first division but in only one of the two cells produced by the second. Natural selection therefore only favours s_d genes if they kill after the second division. Recombination can hit anywhere in the chromosome and a gene does not 'know' whether it is upstream or downstream from a recombinational event. If it kills its sister cell during both cell divisions, or after a random one of the two cell divisions, it will be as likely to kill a copy of itself as of any other gene. Compare with the one-step meiosis in Figure 9.

genes, they can be combined in one body – and totally sterilize it. No life form could last long under the evolutionary attentions of sister-killer genes. Some early life forms, existing soon after the evolution of sex, may have used a simple one-step reproductive cell division; but they would have fallen to bits in a squall of genetic conflict, which is why we no longer see them. We, like all complex life, are descended from the life forms that then evolved mechanisms that neutralized sister-killers. The task, again, is to randomize the information used by the selfish gene.

Figure 10 illustrates the mechanism. The cell initially contains one copy of a sister-killer gene. The first thing the cell does is to double up its genes, making two copies of the sister-killer. The second copy is then used to frustrate the first copy. When the cell divides, one of the sister-killer genes will go in one of the daughter cells. But will the other copy go in the same cell or the sister cell? The answer is effectively a toss-up, and the sister-killer is left uncertain whether its sister cell does or does not contain a copy of itself. A sister-killer gene that kills its sister cell after the first meiotic division has a 50 per cent chance of killing a copy of itself. The Mendelian mechanism achieves the coin-toss in the following way. Initially there was, for each chromosomal pair, one maternal chromosome and one paternal chromosome. They are doubled, making two maternal and two paternal chromosomes. There are then two possible ways of reducing the four copies of each chromosome to one copy per gamete. It could be that, at the first cell division, two of the four chromosomes were picked at random and put in one of the daughter cells, and the other two chromosomes put in the other cell. This would make a sister-killer gene uncertain about whether another copy of itself was in the sister cell. However, the exact coin-toss procedure that we use is slightly different. The two maternal chromosomes (and likewise the two paternal chromosomes) remain attached after the chromosome doubling, and one attached pair of chromosomes goes into one daughter cell, and the other pair into the other daughter cell. The two of the four chromosomes that are put in a daughter cell are not picked at random.

A chromosome is analogous to a book of instructions, with the sister-killer gene as an instruction on one page; it says, 'kill the sister cell after the cell division.' Suppose the book is 100 pages long and the sister-killer is on page 33 of the maternal chromosome. After the chromosome doubling we have two attached maternal books with the sister-killer instruction on page 33 of each, and two attached paternal books with some innocent instruction on page 33. In Mendelian recombination, one of the maternal books and one of the paternal books are split open at a random page – the same page in both books. We can call it page x. Pages $1-x$ are then swapped between the two copies. If x is less than 33, for instance page 10, then pages $1-10$ are swapped and the sister-killers at page 33 are unaffected; they remain attached and both enter the same daughter cell. If x is more than 33, for instance page 40, then pages $1-40$ are swapped between one of the maternal chromosomes and one of the paternal chromosomes. The two copies of the sister-killer gene are separated, and enter different daughter cells. The process works like a toss-up because the chromosomal books are opened and recombined at random sites. A random section of the maternal and paternal books is swapped, and the two sister-killer genes are equally likely to be separated or left together.

The uncertainty extends to the second cell division, when it is again a toss-up whether the second copy of the sister-killer gene will or will not be in its sister cell. Thus, by doubling the gene to begin with, and then using two coin-tosses to decide the fate of the second copy of the sister-killer gene, the information that a sister-killer gene could use in a one-step meiosis has been randomized away. In the one-step meiosis, a sister-killer gene 'knows' it is not in its sister cell; in the two-step meiosis it does not. It no longer knows which other cells do or do not contain copies of itself. Indeed, the process creates maximum uncertainty: any one cell after a meiotic cell division is equally likely to contain a copy of the sister-killer gene, or the alternative victim gene. The doubling-up and random swapping of the genes forces the sister-killers, like the assassin genes, into a game of Russian roulette with $50:50$ odds. That is not a game the gene can gain from, and peace will be restored among the sister cells. Life has become protected from the drawn-out evolutionary tragedy in which, with a one-step meiosis, natural selection crams one sister-killer gene after another into the genome, until there are no gametes left to be killed.

It is essential that the site of recombination is random. If recombination always took place at page 30, then any gene on pages

31–100 could evolve into a sister-killer after the first meiotic cell division (and any gene on pages 1–29 could evolve into a sister-killer after the second meiotic cell division). In fact recombination does open the chromosomal book at a random page. It is one of the unsolved questions of biology how this random choice is achieved. A chromosome is hundreds of millions of letters long. The sites of recombination are not completely random along the chromosome: there are many special regions ('hot spots') where recombination is particularly likely to happen, and the actual sites are probably selected at random from among these hot spots. The final selection of a site for recombination is made by the machinery – an enzyme called an endonuclease – that cuts the strands that are going to be recombined. We need to understand how that enzyme selects its sites. It is a basic requirement of complex life that the selection is done at random; but it is not easy to devise randomizing machines. The fate of the genes depends on the moves made by the endonuclease machine, and if the moves can be predicted or corrupted then selfish genes will evolve to take advantage of it. The machinery has to be blind, but incapable of being led astray. I should like to know how it works.

The theory fits in well with some other facts. Alex Mira argues that it may explain the timing of the meiotic cell divisions in an animal's life. In female mammals, the cells that will later be used as eggs are manufactured early in development, about three to eight months into gestation: the cells that are picked off one a month for ovulation in a thirty-year-old woman have all laid in a state of suspended animation since before she was born. The important fact is that the cells are kept at the $4n$ stage, just before the final two cell divisions. No sister-killer gene can gain an advantage by killing any other cell at the $4n$ stage because every gene (including the sister-killer gene) is in every cell. If the cells were instead kept at the final n stage, sister-killer genes could have thirty or even forty years to practise their dark arts on the other cells of the ovaries. Moreover, the second meiotic cell division is not completed until *after* fertilization. When the sperm enters the egg, the egg still contains two sets of its own genes, though one is in the process of being expelled. The second maternal gene set is not finally expelled until after the sperm is safely in. The risk from sister-killer genes, if they can still act after the confusing gene-spin of meiosis, is lowered still further by holding the cell divisions back to the last minute.

Haig has also looked at some life forms with what he calls 'alternatives to meiosis'. Meiosis in the form we do it (Figure 10) is found in all the large and familiar eukaryotic creatures, but some descendants of the oldest eukaryotic branches have their own idiosyncratic methods of reducing their gene numbers, for purposes of sex. Haig argues that these methods make sense if they evolved to solve the same problem as did our two-step meiosis: foiling the sister-killers. He identified two alternatives to meiosis, in red seaweeds and an obscure group of microbes called microsporidians. One alternative is to multiply the copies of a gene not just to four, as we do, but to a much higher number; in biological terms, they polyploidize like crazy. Then a smaller number of genes are sampled from the big set to form a reproductive cell. The cell might, for instance, multiply its genes to $100n$ – fifty copies of the paternal genes, fifty of the maternal – and then sample a few of them to form each gamete. The information that a sister-killer gene would need is lost in the huge number of copies of each gene. Every gamete has about the same chance of drawing a paternal or a maternal copy of the gene, and a gene in any one gamete has about a 50 per cent chance of being in any other gamete. A second alternative is to go to a $4n$ stage like ours, but then randomly select the two of the four chromosomes that will go to each daughter cell. This is a slightly less orderly method than ours, but achieves the same 50 per cent chance that a gene in one gamete is also in its sister gamete. One final prediction is that one-step meiosis – gamete production by a simple cell division going from a $2n$ cell to an n cell (see Figure 9) – should not exist. This is an eminently refutable claim. Indeed some biologists think it is refuted. But not much research has been done on the creatures that may use a one-step meiosis; the microscopic facts about them are uncertain, and Haig argues that the alleged examples of one-step meioses are unconvincing. The plot will thicken if a convincing example is found, but meanwhile the defence against sister-killer genes is a good explanation for the 'one step back, two steps forward' absurdity of reproductive cell division.

Gene error and gene justice

We have now met a total of three evolutionary advantages for gene shuffling. In Chapter 5 we saw how it can help to concentrate copying errors into a minority of offspring, where natural selection can purge them more powerfully. Here we have seen how it breaks up multigene conspiracies within the body and creates a coin-toss randomizer that prevents the evolution of sister-killer genes. This leaves open the question of what the relation is between these mechanisms. I should guess that sex evolved first to remove passive copying error, and the actively harmful genes that we have thought about in this chapter became a problem after the evolution of sexual reproduction. They are a problem because a gene has only a half-chance of being passed on. In a clonal life form, a gene has a 100 per cent probability of being passed on, and no amount of selfishness can increase this probability. Moreover, when bacteria do use sex they do it in a way that offers no advantage to a sister-killer or an assassin gene. Bacterial sex is unconnected with reproduction and takes place between a donor and a recipient cell: after passing genes to the recipient, the donor bacterium swims off and the recipient incorporates some of the donor's genes into its DNA. Reproduction is a separate event, and when that recipient cell later reproduces, it will simply copy the DNA in its cell and split in two; every gene in the mother cell has the same 100 per cent chance of being passed on to both daughter cells. The chance that a gene in a bacterial cell is passed on when it reproduces is 100 per cent, whether or not the cell has experienced sex. Lawbreaking genes should not have been a problem in the early, simple stages of life, up to the evolution of bacteria and even single-celled eukaryotes. (Other kinds of selfish genes can evolve and disrupt simple life forms, but not the class of genes we are looking at in this chapter.)

Lawbreaking genes became a problem after the chance that a gene was passed on reduced from 100 per cent to 50 per cent. When did that happen? We do not know, but the answer is probably when sex became associated with reproduction. Sex is part of the reproductive process in complex life, but not in bacteria. The change starts to appear in single-celled eukaryotes. Some of them have sex mixed up with reproduction, others do not, and they show an array of meiosis-type gene shuffling processes. Sex becomes more clearly associated with reproduction in multicellular forms of life. A multicellular body cannot take in genes as easily as can a single cell, whether that single cell is a bacterium or an egg. Indeed, no multicellular life form has the ability to replace the genes in one body by the genes from another body. Multicellular life forms like us go through a life cycle in which the adult reduces its cell numbers down to one for purposes of reproduction: down to a single-celled sperm or egg. They can then move genes between cells as easily as bacterial cells can. These gene-receiving cells, however, are also the reproductive cells. Sex has become a reproductive process as the complexity of life has advanced from the single-celled to the many-celled stage. When sex is reproductive, the gene sets have to be halved in reproduction, and there is an ominous reduction from 100 per cent to 50 per cent in the chance that a gene is passed on from parent to offspring. An opportunity was created for a whole new class of disruptively selfish genes. But the pre-existing sexual mechanisms could evolve to keep the peace. Simple life already possessed gene shuffling mechanisms, and the exact form of the gene shuffling could be re-tuned to give it a new peacekeeping role in the new life form. Sister-killer genes were an immediate potential problem when gametes were formed by halving the gene set. Sister-cell killing can be stopped if the gene sets are doubled, and the gene shuffling takes place in the first cell division of a two-step meiosis. Kondrashov's theory of sex is indifferent about whether meiosis has one or two steps, and whether recombination takes place at the first division, the second division, or some mixture of the two. The pre-existing mechanism could therefore be adjusted to its new role without compromising its old error-purging role. The formalities of Mendelian inheritance were needed when sex became associated with reproduction, perhaps when many-celled life evolved from its single-celled ancestors.

A similar story could be told for mate choice. Females might first have evolved an ability to detect gene quality in order to reduce the number of copying mistakes in their offsprings' DNA. But much the same ability will also help against the Mendelian lawbreakers. Gene conspiracies produce their most damaging effects in the sperm (or egg) count, and courtship typically precedes that stage. But the conspirators also reduce the general quality of the body, because of the way genes are coopted from useful to subversive functions. The safe-conduct gene is inferior to the target gene in the absence of an assassin gene, and the evolution of an assassin gene can corrupt a whole region of a chromosome. I suspect a male frog cannot croak so well when he is loaded up with lawbreakers. In the case of the stalk-eyed flies, a conspiratorial set of genes has the effect of shrinking a male's eye-stalks. Females in these crazy-looking creatures prefer males with wide eye-spans, and thus ensure that their offspring are built by peace-loving, cooperative genes. The romance of courtship has been evolutionarily added to the roulette of Mendel's second law, as a further mechanism that enforces Mendel's first law.

But the order in which Mendelian gene shuffling, and courtship, evolved to oppose subversion and copying error in the genes is a historical conjecture. Maybe gene shuffling evolved first against genetic subversion, and became useful against copying error only later, perhaps after life had evolved to be more complex and error rates had gone up. Either way, gene shuffling had a partly accidental role in the history of complex life. It solved one problem and then turned out to solve another. The just severity of Mendel's laws may have accidentally originated in a mechanism to help natural selection clear out copying error, but in their fully evolved form those laws were a necessary precondition for the evolution of complex life.

A prospective feast of Darwinian reason

The two steps of meiosis are not the only intricate and puzzling feature of that interesting cell division. In Chapter 2, when I described some of the other features, I quoted W. D. Hamilton's image of 'this gavotte of chromosomes'. But the dancing partners in the chromosomal gavotte may now appear less innocent than before. They are not debs and debs' delights, or even gigolos and gigolettes. They are like the Montagues and Capulets in *Romeo and Juliet*, or Anna Karenina when she steals Count Vronsky from Kitty, or Don Giovanni in Mozart's opera, where he dances the peasant girl Zerlina off her feet and out through a side-exit at the champagne ball. The drinks are dangerous, the masks confusing, daggers lie concealed in the flowing costume folds. Meiosis is a dance of death as well as love, and your fate may be influenced by your partner, by the surrounding flunkies, or by the music from the orchestra. Its design is still mainly a research programme for the future, but meanwhile we are not short of intriguing facts.

Trisomies, for instance. The cells of a normal offspring, after fertilization, contain two copies of each chromosome. Trisomy means they have three: two copies from one parent and one from the other. A trisomy may concern only one of the chromosomes: in humans, a trisomy for chromosome 21 causes Down's syndrome, a trisomy for chromosome 18 causes Edwards' syndrome, and so on. Or it may concern more than one chromosome, or the whole set. Trisomies cause problems, often major problems, but they are only found in less than 1 per cent of births. The number is much bigger at conception. Many miscarriages in humans – maybe a third of them are due to trisomies. At a conservative estimate, about a third of conceptions are lost before term, implying that more than a tenth of conceptions start with an extra chromosome, a fraction that is ten times bigger than at birth. The true fraction of trisomic fertilized eggs may be even higher; one often-quoted professional estimate is that half our fertilized eggs contain a chromosomal abnormality, and trisomies are a big fraction of them. Trisomies are probably even more frequent among unfertilized gametes. We are not talking about trivialities, at least in human beings. What is going on here?

Trisomies are usually thought of as accidents, and that (I imagine) is what, under interrogation, the extra chromosome would itself say. 'Yes, it was just one of those things. I seem to have held on to my partner too long, we did an extra turn, and somehow we both went right instead of one to the right and the other to the left. Then, to my great surprise, I found I was at the palace feast rather than off on the polar expedition.' Surprising, indeed. One thing to look at is the sex difference in these 'accidents'. In Down's syndrome, for instance, the parent who contributes the extra chromosome 21 is usually the mother. There are actually two genetic causes underlying this trisomy. Geneticists, with their usual flair for instantly intelligible descriptive terms, call them 'non-disjunction' and 'translocation'. Translocation means that a bit of chromosome 21 has duplicated itself onto another chromosome – it can happen in either parent and probably is any old accident. It is non-disjunction that I have under suspicion. It is much more commonly the cause of trisomy and means that, in meiosis, both copies of the chromosome went into one of the daughter cells and none into the other. If we concentrate on the cases of Down's syndrome caused by non-disjunction, they are almost always maternal – the extra chromosome comes from the mother about 95 per cent or more of the time. More than 90 per cent of trisomies for chromosome 18 are also maternal. I have not found data for the other chromosomes, but there is a well-known increase in risk of trisomies with maternal age, and this applies to trisomies in general, suggesting that the sex bias of chromosomes 18 and 21 trisomies is the rule for most chromosomes. About 0.5 per cent of fetuses are trisomic among 35-year-old mothers, 1.6 per cent in 40-year-olds and over 5 per cent in 45-year-olds. The effect of age is suggestive, because the accident is known to happen in one of the final two meiotic cell divisions, not before. We saw above that those divisions occur at the last minute, just before the egg is released for possible fertilization, after the cells have been kept in suspended animation at the $4n$ stage since before birth. Human beings may have particular problems with trisomies because they take so long to breed, as compared with mice or rabbits. Menopause is probably a related phenomenon. It is also a human peculiarity, and its ultimate cause may be that confused chromosomes slowly addle human eggs over time.

The two steps of meiosis produce four cells, and in females only one of the four cells forms a gamete whereas in males all four do. For the cells that may form gametes, ovaries are three times as deadly as testes. In a female, three of the four meiotic cells are destined to die; they shrivel into tiny cells called polar bodies and are disposed of. The mechanism by which only one of the four is selected is another mysterious randomizing device that I should like to know more about. But we can at least think about the selective forces that are at work. Imagine a gene in a female cell that is entering a meiotic division. One way leads to death and total extinction, the other to eternal life (or at least to the chance of eternal life). A lawbreaking gene might be able to fix which side of the cell would become an egg, and which a polar body. Alternatively, it might be able to sense which side was the one to back. If a gene on a chromosome gets any vibes that it is on the wrong side of the cell – the side that will become a polar body – it should sneak across to the other side, and hang in there while the spotlights search for castaway chromosomes. No doubt it will usually be found, and the cell will be shot; but it may sometimes be lucky. For the theory to work, the gene needs information about which half of the cell is destined to become a polar body, and which half will become the egg. We can predict that the ovaries will remove this information if they are able to. They are successful most of the time, because Mendel's law usually prevails. But rare trisomies do happen, perhaps when a chromosome cracks the randomizing system. We saw above that the reproductive cells are held from fetus to adult at the $4n$ stage, before the fateful meiotic divisions. The twenty-five years during which sister-killer genes cannot act are also twenty-five years for the un-Mendelian sneaker genes to sniff around. Forty years would give them longer, and forty-five years longer still. You cannot help wondering about the maternal age effect.

Now we see why trisomies may not be simple accidents. If they are so accidental, how come they accidentally happen thirty times more often in females than in males, while females just coincidentally are the sex that kill three-quarters of the cells in meiosis? Trisomies, so far as we know, usually end up in premature death for the body they are in, and it might seem unlikely that natural selection could favour a rogue castaway chromosome. But trisomies may merely reflect what happens when two lawbreaking chromosomes are inherited together. One of them in a single copy may be able to dance its chromosome partner into the polar body; it will end up in the egg, with the correct number of chromosomes. Natural selection will favour it. Complications arise later, when the chromosome with the smart dance-steps has increased in frequency. Now the lawbreaking chromosome may go through meiosis with a partner like itself, rather than with an innocent chromosome, and a disastrous trisomy is the result. But the lawbreaker can be established at a certain frequency in the population even if it cannot take over. Moreover, even if most trisomic individuals do die, the alternative for a chromosome heading for a polar body is certain death, and any chance of survival, however low, is higher than that. Outside every fat egg there is a thin polar body that tried not to get pushed out.

The argument about trisomies is uncertain. There are other possible explanations for the facts, though I do not find them so likely. I am not sure exactly what is going on, but trisomies have my antennae waving about. In general, meiotic cloak-and-dagger work may lie behind many genetic diseases, and menopause may be a distant echo of the evolutionary division between large female gametes and small male gametes 2000 million years ago. I have concentrated on gene shuffling and on the two steps of meiosis because we have theories for them that are well worked out. But the genes will persistently probe for any weak point in meiosis, and I expect that other (and equally pleasing) theories are waiting to be invented about this protection point for multigene cooperation and historic breakthrough in the rise of complex life on Earth.

Imagining Mendelism

As live complexity increased, sexual reproduction had to evolve at some point. Two parents' genes were combined in one offspring. The problems that this chapter has been concerned with begin when that offspring comes, in turn, to breed. Its gene numbers have to be reduced, almost for reasons of logic. If gene numbers were not reduced, every round of reproduction would double the gene numbers. It would not be long, by one of those geometric multiplication series, before the multiplying genes burst out of the nucleus and exploded through the skin. The Earth would eventually be covered by a ball of genes, accelerating into the universe at the speed of light. For the nuclear genes of Earthly life forms, the reduction in gene number has to be a halving. Each gene then has a 50 per cent chance of being passed on to any one offspring. The problem is that this also means each gene has a 50 per cent chance of *not* being passed on. Natural selection will immediately favour any gene that can improve its reproductive odds. Yet Mendelian justice does prevail and most genes do obey the law of segregation. In this chapter I have aimed to show how the reproductive process is designed to enforce Mendelian justice.

The gene numbers have to be halved, but logic alone would allow many kinds of halving. We might think, for instance, that the halving should be exact. Half the gametes of an organism with an *Aa* gene set would contain *A* and half *a*. Probably the fairest thing the organism could do would be to alternate between producing *a* and *A* gametes: first an *A*, then an *a*, then an *A*: alternate *A a A a A a* ... until it died. But real Mendelian justice is not like this. It is probabilistic: the successive justice of the coin-toss. If the first gamete is *A*, the chance that the next is an *A* is 50 per cent and that it is an *a* is also 50 per cent. We have seen why. If the law of segregation was exact rather than probabilistic, it would supply the genes with exploitable information. If the gametes oscillate successively between *A* and *a*, an *A* gene knows that the next gamete is going to contain *a*. It will then kill the blighter if it can. With the coin-toss, the next gamete has an equal chance of being *A* or *a* and the *A* gene is as likely to kill a copy of itself as of the other gene. This uncertainty is uniquely true of coin-tossing justice: any other way of halving the genes would be vulnerable to sub-version. Biological reproduction, and Mendel's laws, therefore necessarily contain an inherently random element.

A gene lies in a body with another copy of itself and a set of many other genes. The chance that it will be passed on to the next generation is controlled as if by a demon, who takes two decisions. The first is the coin-toss to decide whether the *A* or the *a* gene – the paternal or the maternal copy – will be passed on. The demon then turns to decide what combination of other genes will be sent along with whichever of *A* or *a* it has picked. This could be done by revolving a great casino wheel with about 3300 million slots and 100 or more balls whirling around and settling at random. The balls will determine the sites of gene shuffling. Over evolutionary time, the gene combinations are shuffled and the genes do not know which other genes they will be building a body with. They have to evolve to cooperate with them all. Mendel's demon is a randomizing demon who breaks down the information that a gene would need in order to benefit from acts of selfish destruction. This most efficiently justice-enforcing demon can bend a killer's spear into a boomerang, disperse gene conspiracies and even efface gene identities. Mendel's demon is the executive of gene justice in all complex life, and we should not exist without it.

Mendelian justice is similar to the theory of human justice given by John Rawls in his book *A Theory of Justice*. Rawls argues that human beings will choose justice if they are operating behind a 'veil of ignorance'. Imagine you are going to devise or apply laws, or allocate resources, among a number of people. Maybe you have to divide a pie among five people including yourself. If you know which piece is to be received by which individual you may allocate most or all of the pie to yourself; but if you do not (you are veiled in ignorance) you may divide the pie into equal pieces. The gene shuffling mechanisms we have been thinking about enforce Mendelian justice by drawing a veil of ignorance over the genes. A sister-killer gene, after it has been run through a two-step meiosis, is ignorant about whether a sister cell contains a copy of itself. An assassin gene, after recombination, is ignorant of whether the chromosome it would kill is also the chromosome it is itself sitting on. The gene starts by pointing a gun at its enemy, but after the smoke and mirrors of meiosis it ends up not knowing whether it is pointing its gun at its own head or someone else's.

Moreover, in Mendelian justice, unlike in Rawls's argument about human justice, the veil of ignorance is the means by which the justice is achieved. The veil of ignorance in Rawls's argument is a philosophical reasoning device. Rawls calls it 'purely hypothetical' and 'not intended to explain human conduct'. It is intended to clarify our thinking about justice – about what political principles we should choose if we were in ignorance of whether we or someone else would benefit from these principles. In reality, of course, we are far from ignorant. We are acutely knowledgeable about whether we win or lose from a certain law, or social act. No doubt that is part of the reason why justice (at least in Rawls's sense) is only imperfectly realized in human affairs. Justice, however, is almost perfectly realized within a biological body; perhaps the veil of ignorance really does operate there. Genes do not have a built-in knowledge of their own identity; if a gene shoots, it could easily shoot itself rather than some other gene. The recognition problem is real for them, and they have had to evolve special mechanisms to solve it. The assassin gene that targets a particular sequence on another chromosome, or the sister-killer gene that targets the other cell in the meiotic division, are using an evolved skill to distinguish self from other. Further mechanisms can then evolve to bring them back to justice. Modern genes therefore lie behind two veils. One is that naive genes are ignorant of their own identity; the second is that recombination forces even knowledgeable genes back behind the veil. I see the second process as the closer analogy of Rawls's argument. The probabilistically gene-shuffled veil of ignorance is a counter-intelligence operation that foils the evolutionary attempts of genes to solve the recognition problem. Rawls's mechanism in a way applies more powerfully to genes than it does to human beings. This is just as well. Few of us would welcome justice if it was achieved by randomizing our personal identities.

In Chapter 6 we saw how natural selection could be stopped from working between the genes inside one body by making the genes identical. This does not work for sexually reproduced DNA, however. Here we have seen how the Mendelian mechanism stops natural selection from working within the body, by veiling the genes in ignorance about their identities. What does this tell us about whether Mendelism is a contingent, or a necessary, feature of complex life? Complex life probably has to have sex, but I argued in the previous chapter that gender may be a contingent feature of Earthly life, accidentally derived from the origin of complex life on this planet in a merger event. The arguments of this chapter, about how Mendelian inheritance enforces gene justice, do not depend on the eukaryotic merger. They only depend on sexual reproduction. Something like Mendel's demon, therefore, should be found in all complex life in the Universe. There will have to be a reproductive reduction if not a halving of the number of genes (or the number of the spaced-out equivalents of genes), and the genes will have to be assigned their reproductive fates probabilistically, from behind a veil of ignorance.

8
The long reach of the lawbreaker

A hormonal paradox

Meiosis was the greatest piece of lawmaking in history, and reduced the disruptive genes that threatened the existence of complex life to laboratory curiosities and blackboard postulates. However, we shall now see that a certain class of subtly disruptive genes can, at least in the most complex forms of life, work their way round Mendel's law, creeping through the cracks and interstices in the Mendelian veil of ignorance. The theory is at least amazing, and could even prove mind-blowing, and I should like to begin with a concrete example of the kind of fact that the theory will be concerned with.

Figure 11 is a graph with four lines: look at the lower pair first. One shows the effect on your blood sugar (glucose) levels of drinking some sugar dissolved in water. The blood sugar level rises within a few minutes as the sugar is transported from your stomach to your blood. After half an hour or so it goes down again. The reason is that you secrete the hormone insulin; insulin causes sugar to be taken out of the blood and stored, in muscle for example. (In people with diabetes, who either do not secrete insulin or who do not respond to it, sugar keeps circulating round the body until it is excreted. Diabetic urine smells sweet, which explains, via the decent obscurity of a learned language, the term 'diabetes mellitus', from the Greek words for siphon and honey.) The second line in the picture shows insulin levels: as the blood sugar level goes up after the meal, insulin starts to be released; it drives the sugar level down, after which insulin is switched off again.

The upper two lines of Figure 11 show the effects of consuming an identical sugar drink in a woman who is thirty-eight weeks' pregnant. The top line shows that her blood sugar goes up to a higher level, even though the meal is the same. Arguably this makes sense: the pregnant woman is feeding a fetus too; the fetus is fed from nutrients in the maternal bloodstream; and the higher blood sugar level may be needed to nourish the fetus. The first three lines of the picture are therefore unsurprising. Now we turn to the fourth.

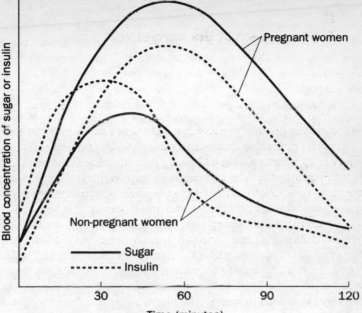

Figure 11. Blood concentrations of sugar (glucose, solid line) and insulin (dotted line) after a meal of glucose, in pregnant and non-pregnant women. The glucose level rises higher and falls more slowly in a pregnant woman, even though she releases more insulin, a hormone that normally lowers blood sugar levels. These are experimental results, and the amount of glucose in the drink is controlled at the same amount in all cases. Likewise, the non-pregnant individuals have been picked to be as comparable as possible with the pregnant individuals – they are similar in age, nutritional state, size and sex to the pregnant subjects.

Blood sugar levels, as we just saw, are controlled by the hormone insulin, and insulin acts to remove sugar from the blood. The obvious way to achieve a higher blood sugar level such as we see in the graph would be for the pregnant woman to secrete less insulin, but in fact insulin secretion is enhanced, not reduced, in the pregnant woman. She achieves a higher blood sugar level despite secreting more of the sugar-lowering insulin. She is responding less to her own insulin (a condition that medics call insulin 'resistance'), which is a rather strange way to increase the blood sugar level. If you showed the two lines for insulin on the graph to most biology students and asked them which would produce the lower blood sugar level (after the same meal, in similar non-diabetic people), they would probably answer that the lower blood sugar level goes with the higher insulin secretion. They would also be wrong.

Why do pregnant women respond less to the insulin that they secrete? To find the answer, we need to look at the work of David Haig, the theoretician whose work underlies most of this and the previous chapter. Haig is an émigré Australian biologist, who now works at Harvard in the USA. He has identified several mechanisms that may be at work in the insulin example. His prime suspect is a hormone released by the placenta and called human placental lactogen (hPL). But before we come to the details of this hormone, it is worth looking at the hormonal powers of the placenta in general.

The placenta is the interface organ between mother and fetus. It is positioned on, and in, the wall of the womb and is connected with the fetus by the umbilical cord. The placenta is developmentally part of the fetus: the cells that make up the placenta are not directly derived from the mother. The fertilized egg gives rise to two structures, the placenta and the fetus itself, which develops into the baby. The placenta receives nutrients from the maternal blood, and passes them on to the developing fetus; waste products from the fetus move in the opposite direction. But the placenta is not simply a passive organ for feeding and cleaning the fetus; it also actively signals the needs of the fetus to the mother, and can probably manipulate the maternal physiology to satisfy them. It does so by secreting hormones into the maternal blood vessels. The human placenta is well positioned to do this because human beings have a so-called 'invasive' placenta. Placentas in different species of mammals have various structures, ranging from kinds that sit on the wall of the womb to our invasive kind in which the placenta physically grows into the womb wall. At the beginning of gestation, cells of the placenta migrate into the wall and surround the maternal blood vessels there. The placental cells partly degrade the blood vessels and replace them with placentally constructed material. We can regulate the blood supply to most parts of our bodies by expanding and contracting the blood vessels, but the mother loses this ability for the blood vessels that supply the fetus. Then the placenta develops endocrine organs – that is, organs for making and secreting hormones – next to the maternal blood vessels. They can secrete hormones directly into the maternal bloodstream, just like the mother's own endocrine organs do; placental hormones do not pass through any kind of maternal filter before entering her blood. Maternal blood contains hormones of both placental and maternal origin, and there is often no way of telling which is which. Placental hormones can influence the mother just as her own hormones can.

Human placental lactogen is one of the hormones secreted by the placenta during pregnancy. It reduces the effect of insulin, and its secretion by the placenta causes the amount of sugar to stay at a higher level in the maternal blood even though the mother is secreting extra insulin. The astonishing thing about hPL is the quantity of the stuff in a pregnant woman's blood. Here are some numbers, compiled by Haig. The concentration of hPL in maternal blood increases to 5–15 micrograms per millilitre at term (a microgram is one millionth of a gram). The concentration in fetal blood is 'much lower': the placenta secretes hPL only into the maternal blood, not into its own. The placenta pumps out 1–3 grams per day of hPL to achieve this concentration. If you are anything like me, these numbers do not mean much by themselves. How do we know that 5–15 micrograms per millilitre is a big number? Maybe most hormones have this concentration. Well, here is a comparison. Human placental lactogen is related to another hormone called human growth hormone. We all use growth hormone, and its concentration gives some idea of what to expect if hPL were any old hormone. Human growth hormone has a daily average blood concentration of 3–6 nanograms per millilitre (a nanogram is 1/1000th of a microgram): hPL has a concentration that is a staggering 1000–2000 times higher than this comparable hormone. Haig describes hPL as 'the most abundant peptide hormone produced by primates'. And if the amounts of hPL are not astonishing enough, it also turns out that hPL is not even needed. Human placental lactogen is lacking in some pregnancies, for instance because the genes for hPL have been deleted by mutation; but the pregnancy proceeds much as usual and the baby is born with a normal birth weight. The mother presumably has other mechanisms that regulate the food supply equally well. Most hormones achieve big effects with a tiny amount of chemical. Human placental lactogen achieves a redundant effect by a relatively large amount of chemical. Nothing in the inherent biochemistry of blood sugar control in human beings

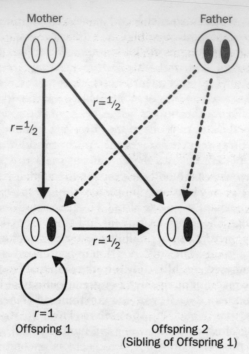

Mother   Father

$r=\frac{1}{2}$

$r=\frac{1}{2}$

$r=\frac{1}{2}$

$r=1$

Offspring 1   Offspring 2
                           (Sibling of Offspring 1)

Figure 12. Natural selection and the parent–offspring relationship. The chance that a gene in one individual is also in another individual is called the relatedness (*r*). Arrows indicate the relatedness from one individual to another. The relatedness between a parent and its offspring is due to direct transfer of genes; that between two siblings is due to the inheritance of the same genes from their parents. A gene in an offspring, such as offspring 1, is definitely in itself (*r* = 1) but there is only a half chance that it is in its sibling. A gene in the mother is equally likely to be in all offspring. Natural selection favours a fair allocation of resources by the mother, but 'unfair' demands by each offspring. The exact calculation follows from Mendel's first law. For siblings, take any gene in one offspring: it came either from the mother or the father. If (chance 1/2) the offspring inherited the gene from the mother, then there is a chance 1/2 that another of her offspring will also inherit it from her, making a total 1/4 chance that the gene is shared with another offspring

is known to require this extravagant and unnecessary output.

So, we have an apparently sensible result (the blood sugar level is elevated in maternal blood); which is produced by paradoxical means (the mother secretes more insulin but responds to it less); which is due to a placental hormone; which is not needed for a normal pregnancy and birth; and which is poured out by the bucketful.

Natural selection goes domestic

Haig saw that a theory of the evolutionary biologist Robert Trivers could be applied to mother–fetus relations. Natural selection, it turns out, works rather differently on genes in the parents and on genes in the fetus (Figure 12). Consider first the genes in the mother that influence maternal behaviour ('behaviour' here includes the way the mother treats the fetus inside her; some may like to substitute the word 'phenotype' for 'behaviour'). Any such gene in the mother has a one-half chance of being inherited in any one of her offspring – the standard chance of inheritance for a gene, as given by Mendel's law of segregation. The same is true of all the mother's offspring; every one of them has the same half chance of inheriting any particular gene in the mother, including any particular gene that influences maternal behaviour.

How will natural selection act on the parental allocation of resources (such as food, protection and other kinds of care) among the offspring? There will be a diminishing returns relation between how much a parent puts into any one offspring and the quality of offspring that results. The offspring benefits mightily from the first few calories that are invested in it, but beyond a certain number of calories the pay-off from further investment levels off. Parents will have evolved to invest a certain amount in each offspring –

via the mother. If (chance 1/2) the offspring inherited the gene from the father, there is a chance between 1/2 and 0 that one of the mother's other offspring will inherit the gene from him, depending on whether it has the same father. The chance this offspring shares a gene with a future offspring via the father is between 1/4 and 0. The total chance of sharing a gene is the sum of the maternal and paternal probabilities, or between 1/2 and 1/4. It is always less than 1.

whatever amount maximizes the number (or number multiplied by quality) of offspring that they can produce. If a parent changed its allocation and put more than the optimal amount into one offspring, that offspring would be of higher quality. But the extra resources allocated to that offspring would have to be taken from other offspring and they would be of lower quality. Their quality goes down more than that of the favoured offspring goes up, because of the shape of the diminishing-returns relation; and the total reproductive output of the parent is less. Natural selection on maternal genes (in Figure 12) should cause resources to be allocated equally among all the mother's offspring. The mother usually will not gain from favouritism, because all the offspring are equally related to her; indeed she will lose from it. Favouritism benefits the mother only if it is favouritism due to differences in need. If one offspring will benefit more than another from some extra care, then that care should be allocated to the needier. It is therefore slightly more accurate to say that natural selection on the mother favours a 'fair' rather than 'equal' resource allocation.

And how about selection on the genes in the fetus? We are thinking about fetal genes that influence the allocation of resources from the mother to the fetus. The gene might, for example, affect a placental mechanism for extracting resources from the mother, or for influencing the level of resources on offer. Does natural selection, acting on these genes, favour the same equal allocation of resources among siblings as it does with maternal genes? The important fact, evolutionarily, is that the gene is certainly in the fetus, but will not certainly be in other offspring produced by the same mother. The exact chance that the gene is in another offspring can be calculated from Figure 12, but for our immediate purposes all that matters is that it is less than 1. Natural selection favours genes in the fetus that place more value on the fetus that they are in than on its siblings. The fetus is not selected to accept a fair allocation of parental resources; it should take (if it can) extra resources for itself.

We can now return to blood sugar, insulin and hPL. Haig argues that we are looking at the result of a past history of conflict between maternal and fetal genes. Sugar would in the past have had some level in the maternal blood. Natural selection on the fetus then favoured mechanisms, such as the secretion of hPL, that increased the sugar supply in the maternal blood. A fetus containing a gene for hPL secretion would have grown up better nourished and had a higher chance of survival; the gene increased in frequency by natural selection. But the fetus with this gene is gaining at the expense of its siblings, and its chance of survival is increased less than its siblings is decreased: the total reproduction achieved by the mother would have been reduced. Selection acts on genes in the mother to counter the hPL, perhaps by secreting more insulin, or by putting out a chemical to destroy hPL. The former, higher reproductive output would be restored. Then, in turn, selection on the fetal genes favours secreting more hPL, to flood the destructive chemical or desensitize the maternal body to the extra insulin. And so on, through millions of years, until we arrive at the megaphone endocrinological diplomacy that seems to be in place now.

Mendelian lawbreaking beyond meiosis

The gene that coded for hPL, when it was spreading by natural selection, was a kind of Mendelian lawbreaker. Mendel's first law states that the two copies of a gene in an individual have equal, 50:50 chances of being passed on. In a species such as the fruit fly, which lacks parental care, a lawbreaking gene can tamper only with the cells that develop into the gametes. It can do so at any stage between when the fate of the cells is fixed and when they finally mature as gametes; for example, the assassin genes that we looked at in Chapter 7 killed early-stage sperm. Later on, assassin-type genes can even bias the chance that one kind of gamete rather than another is successful in fertilization. One gene is known that acts during sperm development and hamstrings the sperm that it is not in, giving its sperm a lead in the subsequent swimming race to the egg. But when the fertilized eggs are sent out into the world the opportunities for Mendelian lawbreaking come to an end. In other species, such as us, the parents care for their offspring after the egg has been fertilized and in them there is an additional phase when lawbreaking genes can act. Mendelian fairness in a species with parental care requires that the two copies of a gene in a parent have equal, 50:50 chances of being passed on when the nest is emptied. But genes in the offspring can act to gain an unfair share of the resources for themselves, such that more than 50 per cent of the final offspring contain the selfish gene. The genes may act any time during parental care – during gestation, or later. They will act by different mechanisms from the classical assassin gene. The assassin gene uses molecular tricks to mark and destroy the gametes that it is not in. A gestationally active lawbreaker gene can work like hPL to influence the maternal physiology. A lawbreaker that acted after birth would use behavioural mechanisms to influence the parental brain, extracting extra resources by appropriately phrased requests. But whatever the mechanism, the lawbreaker gene gains by taking a bigger fraction of a smaller pie. When a fetus first secreted hPL, some of its siblings were short-supplied. Natural selection in the rest of the DNA favours genes that suppress lawbreakers, whenever they act.

In Chapter 7 I argued that the special procedures of meiosis evolutionarily suppress the lawbreaking genes. I have now argued that the gene for hPL is a lawbreaker. Why has meiosis not put a stop to it? The hPL gene resembles a mild sister-killer gene; it is a single gene that is capable by itself of subverting Mendelian justice. Meiosis put a stop to the true sister-killer genes by reflecting the sister-killer's damage back on itself. A sister-killer gene is purely destructive: it kills half the gametes. This can work if the half it kills all contain another gene. But after the genes have been danced through meiosis, a sister-killer gene is as likely to kill copies of itself as of other genes. Natural selection then eliminates the sister-killer genes.

The hPL gene does not simply kill its siblings: it reallocates resources away from them and to itself. It only takes resources from its sibs when it can put the resources to good use. Meiosis again reflects the damage back on the hPL gene and an hPL-bearing fetus takes resources away from other hPL-bearing fetuses half the time. Selfishness still pays, though, because the hPL-bearing fetuses are always the beneficiaries, but make up only half the victims. As a crude analogy, compare burning down a house with robbing it. If you steal someone else's goods, you gain a positive benefit from his or her loss; but burning someone's house down is a simply destructive act – though there could still be reasons why you wanted to do it, for instance to improve the view, or the local housing density. But suppose a demon now blindfolds you and takes you and your incendiary equipment out for a spin, and at the end of the journey, you have a half chance of burning your own house down and a half chance of burning your neighbour's house down. You will probably no longer want to go through with it, now that you have a half chance of self-destruction. Compare that with robbery. Again, a demon blindfolds you and you have a half chance of breaking into your own house and a half chance of breaking into your neighbour's. The demonic veil of ignorance does harm the economics of the project, but does not completely wreck it. You will probably still want to proceed; it is a case of heads you win, tails you do not lose. To make the analogy closer, we could say that there is a chance that the goods are damaged during the robbery, or the house during the break-in. You will still want to proceed, provided that the damage does not exceed half the value of the goods.

We first need to distinguish between lawbreaking genes that are purely destructive and lawbreaking genes that are partly creative. The sister-killer gene, like house-burning, is purely destructive and a two-step meiosis puts a stop to it. The hPL gene, like robbery, is partly creative, and can still be favoured even after a two-step meiosis. In informational terms, with a one-step meiosis, a gene in one cell 'knows' with 100 per cent certainty that it is in that cell and not in the other cell. Vicious genes like sister-killers can evolve. The two-step meiosis alters the probabilities from 100:0 to 100:50. The gene has a 100 per cent chance of being in its own cell, and a 50 per cent chance of being in the other cell produced by the cell division. The genes still have information even after meiosis – the information that the gene is definitely in this fetus but has only a 50 per cent chance of being in another fetus (Figure 12). Mendel's demon has made the genes partly, but not completely, ignorant; the ignorance is enough to limit, but not to stop, selfishness.

Could meiosis be improved on, to hide even the remaining information? I think the answer is that, with sexual reproduction, it cannot. A gene inevitably has 100 per cent information that it is in the cell (or the body) that it is in. The only way a gene can have an equally high chance of being in another cell, or body, during reproduction is if the genes are forced into uniformity, for instance by cloning. This was the conflict-eliminating mechanism that we saw for organelles in Chapter 6. But clonal reproduction is not possible for the nuclear genes of complex life. They use sex, and the Mendelian procedure of meiosis. All that meiosis can do is make the gene maximally uncertain about which other cells (or bodies) do or do not contain copies of it. The maximum uncertainty is reached when every other cell has a 50 per cent chance of having a copy of one gene and a 50 per cent chance of not having a copy of the gene. The informational asymmetry can be reduced from 100:0 to 100:50 but no more. The Mendelian veil of ignorance can only be drawn down so far.

Mendelian justice, therefore, remains open to a subtle kind of subversion. A selfish gene in a fetus is in itself with chance 100 per cent, but in a sibling with chance 50 per cent. If it can obtain, and make use of, resources from its siblings, natural selection can favour it. I suspect that these partly creative kinds of selfish gene, which act after meiosis, would not have evolved until after the origin of parental care. With parental care, there are opportunities for increasing or decreasing the amount of care by small amounts, and reallocating the care among a group of siblings. I find it difficult to imagine the equivalent act in an egg or sperm. We should need something like a 'sister-sucker' gene. A sister-sucker gene is a modification of the sister-killer gene. Instead of killing the other cell produced during cell division, it inserts a pipette-like tentacle into the other cell, sucks out some of the contents, and transfers them to itself. The sister-sucker gene could still be favoured after a two-step meiosis and is closely analogous to the hPL gene. But I do not find it realistic, and suspect that there was little danger from genes of this sort until after the evolution of parental care.

Gene conflict was more than a one-off problem in the evolution of complex life. Sex created a monster problem of gene conflict, which was laid to rest by cunning meiotic proceduralism. But a subtler source of conflict remained, and was probably largely concealed while life continued without parental care. A few Mendelian lawbreakers could be a minor nuisance, but they would be confined to the obscure corners of the DNA where the information-veiling meiotic procedures do not reach. At some point, a form of life evolved that had parental care. It was a step up in the evolution of complexity, because life with parental care is more complex than life without it. But while the care remained rudimentary, it could probably not be exploited or distorted. While the parents simply protected their offspring, and the offspring provided no feedback about their needs, the allocation of resources would remain fair. Over time, parental care evolved more complex forms, including the most intimate and complex form of mammalian gestation. The offspring could now influence their parents, and the parents came to assess their offspring's fluctuating needs by observing signals from the offspring. It was a moment in gene history similar to that moment in human history when a courtier showed the map of the world to Genghis Khan.

A cornucopia of minor lawbreaking

The limit on selfishness imposed by the meiotic gene shuffling means that the selfish genes that act in gestation, or later parental care, will be less damaging than the lawbreakers we thought about in the previous two chapters. But gestationally active selfish genes are likely to be more pervasive. The genes that break Mendel's law before fertilization had a major problem, which I called the recognition problem. A gene inside a cell has no concept of self. If it tries to tamper with the other copy of the gene inside the cell, it has the problem of distinguishing the other copy from itself. A gene holding a gun does not know whether it is pointing it at its own head or at someone else's. The recognition problem could be solved, and assassin genes could spread, only under special conditions. But lawbreaking genes that work after fertilization have no such problem. The information they need is readily available. Conception creates, or soon leads to, an obvious distinction between self and other. The fetus acquires a body as it develops, and a gene can direct resources to itself by the simple expedient of bringing them to the body that it is in. The placenta is umbilically attached to the fetus. There are no difficult recognition problems, as if the umbilical cord might detach itself and whimsically reattach to another sibling or a random member of the population. A concept of self is plumbed-in. A gene expressed in the placenta can automatically help itself by feeding the fetus that it is part of.

This being so, it makes sense that hPL is only one of many paradoxical control systems in mother–fetal relations. Haig's full analysis contains a cornucopia of apparently minor lawbreaking devices, from the immune system, blood pressure and other topics, as well as blood sugar. Consider, for instance, the hormonal mechanism that is used to maintain pregnancy. A pregnancy does not simply persist: it has to be actively maintained, and if the hormones that are used to maintain a pregnancy are not secreted, the pregnancy terminates in a spontaneous miscarriage. Many pregnancies do terminate prematurely in this way. The losses mainly occur in the first twelve weeks of pregnancy, with many in the first few days. The reasons are only partly understood, but one reason – and the one that matters here – is quality control. If a fetus is of low quality, for instance because it contains poorly copied genes, it is more likely to be miscarried. We saw some evidence in Chapter 7, for fetuses with the wrong number of chromosomes: fetuses with chromosome abnormalities are disproportionately likely to be miscarried. The miscarriages are evolutionarily advantageous for the mother, because she can reallocate her efforts to a future offspring, which can be expected to be of average quality. But they may not be advantageous for a gene in the fetus. The genes are definitely in this fetus, but there is only a half chance they will be in a future sibling. Genes in the mother may be in conflict with genes in the fetus about whether to maintain a pregnancy.

The ancestral mechanism of pregnancy maintenance in mammals, and one that is still retained at least as a back-up in human beings, is for the anterior pituitary gland in the brain to secrete luteinizing hormone (LH), which travels in the blood to the corpus luteum in the ovaries, where it stimulates the secretion of progesterone, which keeps the pregnancy going (Figure 13). A mother could in ancestral times have terminated an inferior fetus by switching off the progesterone. But there would then be a selective advantage to a fetus that could keep the progesterone flowing. In modern humans, it is the placenta, not the mother, that has become the main source both of 'luteinizing hormone' and of progesterone during pregnancy. The placenta secretes a hormone called human chorionic gonadotrophin (hCG). Human chorionic gonadotrophin is a molecular mimic of LH and binds the LH receptor in the corpus luteum, stimulating the mother to produce progesterone just as her own LH does. The placenta also secretes progesterone directly into the maternal blood.

The concentrations of the fetal hormones are again enormous. The peak concentration of hCG is 50 units per millilitre at weeks 8–12 of pregnancy, which can be compared with the peak LH concentration, when it spikes up at ovulation, of 0.1 unit per millilitre. Fetal hCG is produced in 500 times the quantity of maternal LH. Human chorionic gonadotrophin is secreted in such large amounts that it is used in many pregnancy tests. To secrete

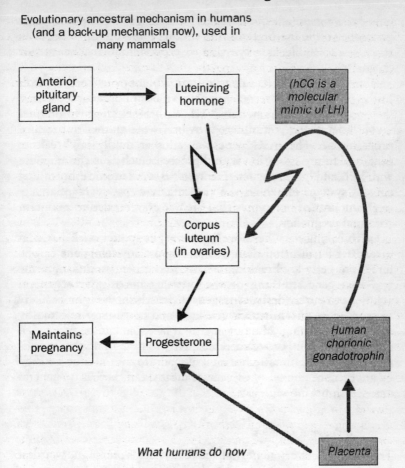

Figure 13. The hormonal maintenance of pregnancy. Pregnancy is maintained by progesterone. The ancestral mammalian system is for the mother's ovaries to secrete progesterone in response to luteinizing hormone (LH). In modern humans, the fetus has taken over, by directly secreting progesterone and by secreting human chorionic gonadotrophin (hCG), a mimic of LH.

hCG is one of the fetus's earliest metabolic acts, and enough of it is produced for it to be detectable in the mother's urine. The placenta also puts out an apparently excessive amount of progesterone. The mother's corpus luteum, left to its own devices, puts out a maximum of about 40–50 milligrams per day of progesterone, while the placenta is secreting more than five times as much – 250 milligrams per day of progesterone by the end of pregnancy.

It looks as if the placenta has, during evolution, usurped control. In the past, inferior fetuses might have been miscarried because their progesterone was switched off. Natural selection would then favour a fetus that restimulated (by hCG) the mother to produce progesterone. The hCG gene was another mildly lawbreaking gene, and future evolution would proceed much as in the example with hPL and blood sugars. The result is a paradoxical hormonal control system, in which the placenta now produces relatively huge amounts of hormones that are not even needed to maintain a normal pregnancy.

Lawbreaking may be more pervasive, in species with parental care, after fertilization than before. Almost any fetal gene can be lured into petty lawbreaking. There is no general solution, in the way that gene shuffling or the two-step meiosis were general solutions to purely destructive genes. The selfishness is more likely to crop up persistently at a low level and be counter-selected by one-to-one policing. Human placental lactogen can be met by an increase in insulin, or reduced sensitivity to hPL, or a specific anti-hPL chemical. Human chorionic gonadotrophin can be resisted by another specific set of molecular means. But there is no general anti-selfish-hormone device.

Imagining conflict

The conflict between maternal and fetal genes is one of the weirdest ideas in the modern theory of evolution. The conflict is not difficult to see: for any marginal calorie, the fetus will often gain if that calorie is allocated to it, whereas the mother will often benefit by allocating it to another (or storing it for a future) offspring. But strictly speaking the conflict is not between the mother and the fetus. It is actually a conflict between maternal and fetal genes, both sets of which are present in both individuals. We considered how natural selection acts on genes in the mother that influence maternal behaviour, and on genes in the fetus that influence fetal behaviour. But the mother also contains a full set of genes that code for fetal behaviour, and the fetus contains a set of genes for maternal behaviour – it is just that they are switched off. Genes in the mother that code for fetal properties (such as the gene for hPL) were expressed many years ago when she was a fetus, but have been silent ever since. Genes for maternal skills such as adjusting blood sugar levels after a meal during pregnancy are present in the fetus but are silent; they may or may not fire into action in the future.

The conflict is between two sets of genes within each individual. One set of genes is turned on at one stage of life; the other set of genes is turned on at another stage of life. Natural selection works differently on the two gene sets at different stages of life. A gene like the one coding for hPL, which gives you an advantage as a fetus, turns disadvantageous when you come to reproduce and your offspring spring it on you. Selection now favours genes that counter the effect of hPL, for instance by increasing insulin production. The hPL gene is like an assassin gene that distorts its parent's reproduction, not its own body's reproduction. The counter hPL gene is turned on in the mother, but exerts its suppressive effect on a gene that is turned on in the fetus. And to cap it all, the two bodies, the mother and the fetus, which have different parts of their common gene sets turned on and off, are conceptually difficult to separate and are parts of one physically continuous whole. As Dorothy would say, 'Toto, I don't think we are in Kansas any more.'

The organismal expression of the gene conflict is scientifically interesting. It shows how the same evolutionary logic applies to the gene conflicts described here and in the previous chapter. But it also matters politically. Mother–fetus relations occupy a sensitive point in sexual politics and Haig's ideas have come under the political gaze. Three points are worth keeping in mind. One is the argument we have just looked at. The conflict is genetic, between genes all of which are in both mother and fetus (and in men, for that matter). It is not an organismic conflict, like a fight between two creatures over a limited resource such as food or territory: the creature that wins the food is unambiguously better off. The fetus with an hPL gene, which can extract extra resources from the mother, does gain while it is a fetus. But it will lose out when it grows up and reproduces. It will have offspring who shrink its own reproductive output; it will produce fewer offspring than average. The fetus, over its full lifetime, loses out by possessing the hPL gene. It suffers much like a fruit fly that contains the assassin gene: the fruit fly will produce fewer offspring, but the assassin gene still spreads because a disproportionate fraction of the offspring contain the gene. The organism with the selfish fetal gene also has fewer offspring, and the gene spreads because a disproportionate fraction of these offspring contain the hPL gene.

Second, the conflict is evolutionary, and may or may not be producing inefficiencies now. The theory we looked at aims to explain the maternal–fetal relation in terms of evolutionary history. We began with an apparently paradoxical control mechanism. We explained it by a series of evolutionary steps, in which the placenta increased its output of a hormone, and the mother decreased the influence of that hormone. The end-point may be much like the beginning, as if everyone in a crowd stands on their toes to try to see further. The extravagant hormone secretion, and the theory to explain it, do not imply that the fetus and mother are manipulating each other now, or that the allocation of resources among offspring is inefficient now. They do imply that some physiological process equivalent to unconscious manipulation operated in the past, and that the allocation of resources was inefficient in the past. Haig's work concerns various features of maternal–fetal relations that would appear odd if you did not know about the gene conflicts in their evolutionary history. The evolutionary conditions for the conflict do continue to exist, and the next section looks at what effect they may be having now.

Finally, we have been concerned with physiological, not emotional, mechanisms. The theory does not imply that mothers do not love their babies; indeed if it did that would be a good reason to reject it. In general, we see emotional conflicts in many human, and non-human, relationships, including parent–offspring relationships. The conflicts produce attention-grabbing results, such as (depending on the relationship) shouting matches, punch-ups, lawsuits and shoot-outs. We see people posturing and pulling faces, or performing behaviour patterns like the one known in grown-up politics as 'throwing teddy round the nursery'. We also sometimes see nothing at all, for a while – followed by a sudden mean trick or an apparently spontaneous flare-up. I should like to be able to explain these behaviour patterns, but it would take further work before we could say anything about them similar to what we can say about escalated hormonal signalling. We should need to know, for instance, the relation between emotional display and resource allocation. I think it is currently uncertain how much gene conflicts have to do with these sorts of behavioural and emotional conflicts. Gene conflicts, by contrast, almost certainly have a great deal to do with paradoxical hormonal signals.*

The Hippocratic gene

Do conflicts between genes reduce the efficiency of life? The answer depends on what evolutionary result the conflict leads to. The conflicts that we looked at in Chapters 6 and 7 led to the evolution of a general conflict-resolution mechanism that put a permanent stop to the selfish genes. In this case, life is probably almost as efficient as if the conflict had never existed. Gender does perhaps produce some inefficiencies. It halves the rate at which individuals encounter possible mates. But mate encounters are not the rate-limiting step in breeding, and I doubt whether the rate of pro-creation would differ if we were a de-gendered, unisex human-equivalent species like the one we thought about in Chapter 6. Some other species are rarer than us, and may have more of a problem in meeting mates, but they have usually solved the problem by evolving hermaphroditism. A hermaphrodite can mate with any other individual of its species, even itself if it is lonely enough. (Woody Allen jests that bisexuality doubles your chance of a date on Saturday night. Someone should tell him about hermaphroditism.) Gender can also reduce the scope for mate choice, because the genetically best individual you meet may be the same gender as you. In all, the evolution of gender resolved a gene conflict and then produced either trivial or minor inef-ficiencies.

The paradoxical hormonal control mechanisms that Haig has

* In pedantic terms, we can distinguish three kinds of conflict: (i) conflicting selection pressures on the genes within the DNA of one organism; (ii) past physiological conflicts that result in escalated hormonal signalling; and (iii) modern emotional conflicts. I have argued, following Haig, that (i) explains (ii). The explanation of (iii) is an interesting question, but more difficult than (ii). I have made no claim about whether (iii) is or is not analogous to (ii) and also explained by (i). Also, any difficulties that someone finds in explaining (iii) are not necessarily a reason to doubt (i) or its explanation for (ii).

identified are another case. No general conflict-prevention mechanism has evolved, and individual lawbreaking genes have evolved from time to time. We can think about their effect in two stages. The first is while the gene (such as the gene for hPL) first evolved. It initially biased the mother's allocation of resources away from the optimum and reduced her reproductive output. At this stage it was reducing the efficiency of life. Later on – after a few thousand generations – the hPL gene had spread and was present in everyone. Other genes had evolved, to enable the mother to ignore or counteract hPL, and the maximally efficient *status quo ante* was restored. All that was left was the minor inefficiency of synthesizing a few grams of an unnecessary hormone. There was a temporary inefficiency while the lawbreaking gene swept through the species, but little or no permanent inefficiency.

We do not know how frequently anti-Mendelian genes such as hPL arise, and how much they reduce efficiency for the average human being during evolutionary history. Good evidence exists for paradoxical fetal control mechanisms that have evolved in the past, but these may no longer be producing an inefficient resource allocation. The evidence we have does not therefore demonstrate that modern mothers allocate resources anything less than optimally. But the potential gene conflict is always present. It is theoretically possible that we contain many rare genes with sneaky tricks for extracting, or blocking, resources, and that they may bias the resource allocation from time to time, depending on the details of an individual mother and fetus. I prefer to keep an open mind about whether genes of this kind are at work.

Another possibility is that the gene conflict is not now causing resources to be allocated inefficiently, but the resultant control mechanisms have led to other medical problems. Let's stay with the blood sugar example. As we saw, mothers routinely become partly insensitive to their own insulin during pregnancy. This insensitivity comes in various degrees and we may be looking at the thin (or not-so-thin) end of a wedge to diabetes. You can be slightly unresponsive to insulin, moderately unresponsive, or seriously unresponsive; at some point in this continuum you are clinically defined as diabetic. What happens in pregnancy is that the mother becomes slightly diabetic – and in 10 per cent of pregnancies she becomes sufficiently unresponsive to insulin to be defined as having gestational diabetes. So far, no problem:

gestational diabetes is on average a good diagnosis for the pregnancy, and it disappears after birth. But there are some suggestive additional facts. Gestational diabetes is followed in 50 per cent of cases by full diabetes later in life. Mechanisms such as hPL have evolved that tend to make mothers diabetic in pregnancy, and the evolutionary conflict over blood sugar regulation may have predisposed us, or some of us, to a problem with diabetes. Diabetes can be treated, but this does not mean we can relax about it. Diabetes is a global health problem, a major cause of blindness and death. Many other things besides pregnancy influence your chance of being struck by it: there are well-known risk factors in diet and physical exercise (or rather, lack of it). The percentages above were for rich Western populations, which are famous for their evolutionarily egregious diets and exercising habits: they may exaggerate the evolutionary, if not the modern medical, importance of gestational diabetes.

If diabetes is in part a consequence of maternal–fetal gene conflicts, then these conflicts have led to a more permanent inefficiency in human biology. The inefficiency is not the same as a misallocation of resources. In an escalating shouting match, an individual may gain more of the resources by shouting louder. Resource allocation is inefficient for a while. Then people re-adjust their auditory thresholds, and the allocation is restored to the optimum. The process can then repeat itself until it results in 'megaphone diplomacy': everyone shouts their heads off, ignores everyone else, and resources are efficiently allocated. But the final state may harm people in other ways: they may be hoarse, or have ringing ears. The evidence for diabetes also hints that the conflict has produced an inefficiency, though the inefficiency is not in the resource allocation. A full argument would be even more complicated, because a life form that was equivalent to us but lacked a history of gene conflicts might have problems of its own. A humanoid with an ancestry of virgin birth would lack gene conflicts, and might be less diabetic than us. But its genetic set-up might lead to problems of its own, even ignoring the difficulties with harmful mutations that we saw in Chapter 5. I do not know what the conclusion of the argument would be. In any case, the medical implications of Haig's ideas are mainly a research topic for the future. It is like one of those maps with a cross that marks the buried treasure. When you visit the site, the scoffers will declare that all is darkness and go home. But if you shine a light down the mineshaft, it flashes back, sparkling in the distance. You can tell that the answers are lying down there, like an ancient hoard of rubies and diamonds, jade and gold.

More fruit from the tree of knowledge

So far in this chapter I have only assumed that genes have information about whether they are in a fetal or a parental body. Natural selection then favours one kind of behaviour in fetal genes and another in maternal genes. The assumption is certainly reasonable and almost certainly valid. All that it requires is that there is some mechanism that can switch genes on or off depending on whether they are in a fetus or a mother – and mechanisms of this kind surely must exist, because mothers and fetuses differ in so many ways. This does not mean that the argument about blood sugar is certainly true, but it does mean that it is built on a reliable assumption. We can now advance the argument by thinking how natural selection might act on some more subtle (and less obviously available) kinds of information. The ideas are again all due to Haig.

The first possibility is that a gene in the mother may be able to recognize copies of itself in the fetus, or vice versa. There are two copies of every gene in the mother; we can symbolize any one pair of genes by Aa. A fetus inside her has a half chance of inheriting her A gene and a half chance of inheriting her a gene. Suppose that it is a gene that is turned on in the mother, and influences the transfer of resources to the fetus. We have been considering the case in which the gene has an equal chance of being in any of her offspring. Natural selection then favours a version of the gene that allocates resources fairly among all the offspring. But now suppose that an A gene in the mother can recognize whether or not it is in a particular fetus. The half chance that an A gene is passed on is an average for half the offspring who definitely do have A and half who definitely do not have A. If the A gene in the mother can recognize which fetuses do, and which do not, have copies of itself, then natural selection will cause it to allocate all the maternal resources to the A-bearing fetuses and none to the a-bearing fetuses. This information – the information about which fetuses contain the A gene – opens up an opportunity for a bizarre kind of cross-generational Mendelian lawbreaking. The gene enjoys the 100 : 0 certainty of the unrestrained lawbreaker rather than the 100 : 50 uncertainty of Mendelism.

How could a gene recognize copies of itself? It might do so among the cells, and molecules, at the maternal–fetal interface in the womb. Certain cells in the mother and the placenta have recognition molecules on their exteriors. A recognition molecule can bind another molecule that has the appropriate shape, and the behaviour of the cell may then change. For instance, when the recognition molecule is bound it may start a cascade of biochemical reactions inside the cell, and the cell may come to inhibit, or promote, fetal growth. For our purposes, the interesting recognition molecules are self-recognition molecules: ones that bind other molecules that are the same as themselves. Recognition molecules may, depending on which one we are talking about, bind different molecules – that is what antibodies do, for example – or the same molecule as itself. We concentrate here on the latter. One example is a set of molecules called cadherins. Cadherins cause cells to stick together, which is a useful skill in the body because we need common groups of cells to stick together at various times and places. Cells with cadherins on their exteriors are found in both the placenta and the surrounding maternal parts of the womb. Now think of the genes for any one of these self-recognition molecules in the mother. She contains the usual two copies of the gene (she is Aa), but puts only one of them in the fetus. To make the argument concrete, imagine that only one of the two genes in the mother is expressed on certain of her cells; the A gene, for example, might produce a protein that sits on the surface of certain cells. Imagine also that the placenta contains cells, expressing the copy of the self-recognition gene they inherited from the mother; if it inherited A it has A-bearing cells, and if it inherited a it has a-bearing cells. We now have the potential for a lawbreaker gene that can recognize other copies of itself.

What might happen? We can imagine that these maternal cells circulate round the body hopping on and off the bloodstream. Some stop by at the placenta, and wander about, shaking their molecular hands with the local placental cells. If the A-bearing maternal cells encounter A-bearing placental cells, they 'know' that the fetus inherited the gene they express rather than the other gene. The successful molecular self-recognition raises the maternal cell's estimate that the gene is in the fetus from 50 per cent to 100 per cent. A fetus that certainly contains the *A* gene is more valuable than average. It deserves extra care, extra nourishment, extra protection from the maternal immune system. The encounter between the cells might therefore stimulate the maternal cell to have extra resources directed to the fetus. As usual, the advantage of the selfish gene, the selfish self-recognition gene in this case, reduces the total reproductive output of the mother. It is a disadvantage for the other genes in the body and they are selected to suppress it; but there may still be opportunities for a gene that can recognize copies of itself. Self-recognition molecules potentially are not fooled by the randomizing tricks of Mendel's demon, and they may have unusual opportunities to help themselves at the expense of the rest of the body. They are known to exist, and on cells that operate in the maternal–fetal relationship; but we do not know what they do there, or how they evolved. All we have for now are the reports of biologists who may have unwittingly glimpsed the Byzantine politics of the self-recognition molecules.

The second kind of information is less hypothetical. It arises when the expression of a gene depends on whether it was inherited from the mother or the father: this is known to happen in some genes and it is called genomic (or genetic) imprinting. An 'imprinted' gene might be silent if it was inherited from the father but expressed if it was inherited from the mother, or vice versa. Most of the time it makes no difference because the two copies of the gene are the same, and the body needs only one of them to be active; but the results are strangely non-Mendelian if the individual inherits different versions of the gene from the two parents. In a standard Mendelian cross, such as the cross between pea plants with pure purple (*PP*) and pure white (*pp*) flowers, the offspring are all purple and have a *Pp* set of genes. If these genes had been imprinted, the colour of the offspring would depend on whether the mother or the father was purple. If the genes concerned were paternally active, then a *Pp* pea plant would have purple flowers if its father was purple and its mother was white, but white flowers if its mother was purple and its father was white. Mendel had enough to puzzle out with the crosses he did, and he can count himself lucky (and we can count ourselves lucky) that he did not pick on any imprinted genes. Genomic imprinting was discovered much later.

The activity of imprinted genes depends on whether they came from the father or the mother of the body they are in. It is as if the parents mark some of their genes before putting them in a gamete; a marked gene may be switched permanently off in the offspring, or left to be switched on as appropriate, like a normal gene. What matters for us is the information that the marking supplies. An imprinted gene in the fetus now 'knows' whether it is maternal or paternal. Natural selection may set to work on the genes, depending on the conditions. One condition in which natural selection will set to work is in a promiscuous species in which the successive offspring of a female are fathered by different males. Consider the maternal and paternal copies of a gene in any one offspring. Natural selection acts on the maternal copy of the gene in the way we thought about earlier in the chapter. The maternal copy has a 50 per cent chance of being in a future offspring, and it values the offspring it is in twice as much as a future sibling. But the paternal gene in the offspring has a 0 per cent chance of being in the mother and a 0 per cent chance of being in her future offspring. Natural selection on it will favour unrestrained resource-gouging, working the mother to death, having her risk her life to provide a final 0.000001 per cent increment to the quality of the offspring. The paternal gene places no evolutionary value on the mother's future reproduction.

The theory is supported by several examples. Real fetal genes are known that tend to encourage fetal growth, and it turns out that they act only when they are inherited from the father, not the mother. Other fetal genes tend to limit the growth of the fetus, and they act only when they are inherited from the mother, not from the father. An early example, discovered in the 1980s, concerns two sets of genes in mice. One gene is a growth factor, and the paternal copy of the gene is active in the fetus while the maternal copy is silent. The other gene seems to code for a molecule that binds to and degrades the growth factor, and the maternal copy of this gene is active while the paternal copy keeps quiet. The pattern makes sense: mice are not long-term monogamists, and the paternal genes in a fetus value this fetus more highly, relative to future offspring, than the maternal genes in the same fetus.

At this point we are still on solid ground. The theory and facts are both reasonably reliable, and the former probably explains the latter. Conflict of this kind between paternal and maternal genes have probably limited the degree, and influenced the form, of complexity in life. I want to finish by peering as if through a telescope beyond the frontiers of scientific knowledge and into unexplored territory. The complexity of life has increased not only through parent–offspring relations, but also through social relations. In a complex society, individuals other than parents and offspring cooperate and divide up the labour. We see social cooperation in ants, bees, monkeys and many kinds of birds and mammals. Female mice, for instance, may help their female relations to bring up their families, presumably at some cost to their own reproduction. Cooperation typically evolves between genetic relatives, such as siblings or cousins. An individual can propagate genes inside itself by assisting the reproduction of other individuals – genetic relatives – who also contain copies of the same genes.

Now for the fireworks. In mice, we know something about which genes are expressed in different developing regions of the brain. Genes from the father are more active in the hypothalamus and genes from the mother are more active in the cortex and corpus striatum. The paternal genes that work in the hypothalamus influence (among other things) the maternal behaviour of the mouse, if it is a female mouse. Female mice that lack a paternal copy of a gene called *Peg3* that is active in the hypothalamus (and elsewhere in the brain) make bad mouse mothers. They are slow to build a nest, slow to retrieve wandering pups, slow by other measures; their pups grow about 30 per cent less than normal mice pups. Less is known about the maternally inherited genes that are active in the cortex and corpus striatum, but they are suspected of influencing social behaviour. Simplifying and going beyond the evidence, it is as if the paternal genes in a female mouse direct her reproduction and the maternal genes direct her social relations.

Why should paternal genes promote motherhood in a mouse? We do not know for sure but we can think about it. Look at Figure 14. The important feature in the figure is that the successive offspring of a female have different fathers. A paternal gene in one offspring is unlikely to be in its siblings. The paternal gene can be propagated if the offspring breeds itself, but not if it helps its siblings to breed. The maternal gene, by contrast, is in all the siblings, and may be propagated more efficiently by a daughter who helps her sisters rather than reproducing herself. The paternal

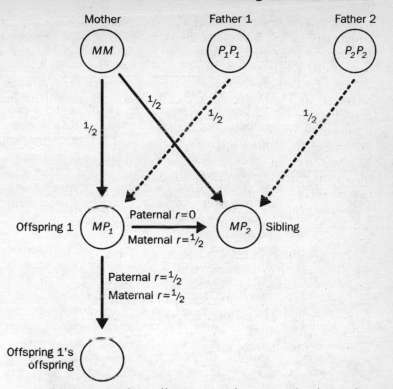

Figure 14. Look at offspring 1, and compare the chance that a gene in its father is also present in (i) offspring 1's sibling or (ii) offspring 1's offspring. The chance that it is in the sibling is lower, because offspring 1 and the sib are fathered by different males. But for a gene in offspring 1's mother, the chances are the other way round. A gene in offspring 1's mother has a higher chance of being in one of offspring 1's siblings than in one of offspring 1's offspring. The sibling is the mother's offspring whereas offspring 1's offspring is her grandchild. A paternal gene in offspring 1 may relatively favour breeding over social cooperation, and a maternal gene in offspring 1 may favour social cooperation over breeding. There is evidence that brain cells in mice, in regions of the brain that control maternal behaviour, develop under the influence of paternal genes. Social behaviour may develop more under the influence of maternal genes. For the meaning of the terms and arrows in the picture, see Figure 12.

and maternal genes inside any one offspring differ in the value that they place on reproduction as opposed to helping a sibling, on breeding as opposed to social cooperation. And the different values of the paternal and maternal genes seem to match their brain activity. It is tempting to infer that, over evolutionary time, the paternal genes have seized control of the breeder hypothalamic cells, and the maternal genes have seized control of the social cortical cells. A developing mouse brain is wired up by two electricians, one hired by the mouse's father and the other by her mother. The spheres of influence of the two electricians seem to make sense, given the evolutionary interests of the father and mother of the mouse. There is the possibility here of a unified evolutionary explanation, which connects social behaviour with the development of brain cells, and with the peculiar post-Mendelian information provided by imprinting.

Mendelian inheritance places the genes behind a sexual veil of ignorance. In simple Mendelism, a gene does not know whether it came from the mother or the father of the organism it is in. This ignorance imposes a certain justice on the genes, limiting the selfishness that will evolve. But the game changes when a gene knows whether it is paternal or maternal. The inference about mice is speculative, and close to the frontiers of science; it may prove completely bogus. But the underlying principle still matters. Male and female behaviour differs in all species, and this will often result in different patterns of similarity between the paternal and the maternal genes of two individuals. The fact of genomic imprinting shows that the genes can know the sex of the parent they came from, and that natural selection has set to work accordingly. The veil of sexual ignorance, which Mendelian inheritance had once placed on our genes, has been torn away.

Gene conflict and live complexity

A living creature can be harmed by passive copying errors in its genes, or by subversive genes that are actively favoured by natural selection. Chapters 3 to 5 were about passively harmful error; Chapters 6 to 8 have been about active subversion. Actively subversive genes are likely to spread whenever the conditions exist for gene conflicts: when one gene can gain at the expense of other genes in the body. A simple life form can cope with relatively anarchic relations between its genes; it does not require the cooperation of large numbers of genes. But large numbers of genes do have to cooperate, to build each body, in complex forms of life. Complex life has to be structured so that genes, or at least most genes most of the time, do cooperate. The danger from conflict shot up about the time of the origin of the eukaryotic cell, and of Mendelian sex, over 2000 million years ago. They eukaryotic cell originated in a merger between at least two kinds of cell; the resulting cell would have contained at least two independently replicating DNA molecules. Sexual reproduction also puts two independently replicating DNA molecules in one cell, and a sexually reproducing cell after the eukaryotic merger could have had four replicating DNA molecules.

We saw in Chapters 6 and 7 how these four DNA molecules have been forced to cooperate. Gender does part of the job: the males dispose of their organelles before breeding, which reduces the number of DNA molecules in an offspring cell from four to three. The flight of genes from the organelles to the nucleus has also helped; the nucleus is now so dominant that the organelle genes have few opportunities for selfishness. That leaves two: the male and female nuclear genes. The job is completed by the gene shuffle of meiosis, which randomizes away the information that either of these nuclear DNA molecules would need to execute any acts of selfishness. Gene justice is achieved by unification, and by the veil of ignorance.

Gender, gene transfers and meiosis did not evolve to cause the evolution of complex life. Evolution is not capable of this kind of strategic planning. Indeed some of the gene shuffling procedures had probably already evolved, to purge copying errors; they were then re-tuned to enforce the Mendelian law. Once meiosis, gene transfers and gender had evolved, they in turn enabled complex life to evolve, and as we look backwards from forms that are complex we can see that they could evolve only in lines that somehow solved the problem of gene conflict. Gender, gene transfer and meiosis have in common a conflict-resolution device that prevents future conflict from flaring up. They are like a human cultural rule that pre-empts the possibility of conflict. The rules about rights of way on the road are a response to a conflict, and also effectively prevent (most of the time) any observable manifestation of the conflict.

Meiosis, gene transfer and gender are all connected. They form the basic set-up of Mendelian reproduction. Almost all the genes of complex life forms are inherited in a Mendelian manner, and a description of Mendelian inheritance is implicitly for genes in the nucleus, in a sexual system with fair inheritance and random gene shuffling. Moreover, the same system works against both the actively harmful lawbreaking genes and the passively harmful copying errors. I therefore see the evolution of the system – which I am calling Mendel's demon – as the great breakthrough in the rise of complex life on Earth. Other anti-error devices had evolved both before and after the demon – devices such as repair enzymes and mate choice. But the pre-demonic devices took life only as far as bacteria, and mate choice depends on the prior evolution of Mendelian sex. Mendel's demon itself is the key to the creation of complex life on Earth.

I described, in Chapter 1, two big facts about the history of complex life. One is that all complex life is built of eukaryotic cells. What, I wondered, is it about the eukaryotic cell that is the secret of complexity? Eukaryotic cells differ from bacterial prokaryotic cells in many ways, and the full set of eukaryotic properties probably evolved over hundreds of millions of years. Some eukaryotic properties are probably irrelevant to the evolution of complexity; others are probably crucial. We are now in a position to guess which is the most crucial factor of all: it is Mendel's demon, the set of procedures that control the inheritance of genes.

Mendel's demon is a uniquely eukaryotic possession. Bacteria and other simple life forms mainly clone themselves, though some of them do occasionally use certain kinds of mock-sexual gene shuffling. But they all lack the special cell division of meiosis, and the kind of systematic, fair-exchange, gene-shuffling, gendered sex that we and other complex life forms use. I do not know when the demon evolved during the history of the eukaryotic cell. It must have been some time after the origin of the eukaryotic cell, because not all eukaryotes have meiosis as we do. We saw in Chapter 7 a reason to think that the full demon would have evolved when sex became connected with reproduction, and that may have finalized in the evolution of many-celled life. Nor am I sure that the demon had to evolve in a merged-cell life form, though some demonological details, such as gender, do depend on an ancestral merger.

But the demon did in fact evolve at some stage in a eukaryotic cell, and only in a eukaryotic cell. Its evolution opened up the possibility of future complex life, and it is the reason why all complex life is built of eukaryotic cells. The other big fact about the history of complex life is its long delay while Earth was dominated by simple microbes. We saw that the dates for the origin of the eukaryotic cell and of many-celled life are sufficiently uncertain that we can read history in two ways. One possibility is that the delay may have been over the eukaryotic cell. If so, the delay could reflect some improbability in the merger that created the eukaryotic cell, and the demon may then have followed relatively easily after this improbable event. Alternatively, the eukaryotic cell may have evolved more rapidly and the big delay was in the evolution of many-celled, rather than single-celled, eukaryotic life. Then the delay could reflect a difficulty in the evolution of the demon itself – it is, after all, a complex and elaborate set of procedures.

Mendel's demon can concentrate error, resolve conflicts and randomize information effectively enough to allow the evolution of creatures like us, who may live for a hundred years with 60,000 gene instructions and 6600 million letters of DNA code. The demon acts to bring out the best in the genes, evolutionarily encouraging them to cooperate for the good of the whole body. But it is still only effective up to a point. Meiosis increases the uncertainty about which other gametes, or offspring, contain copies of a gene. It has stopped, or almost stopped, the evolution of the most damaging kinds of sister-killing and assassinating genes. The genes lived harmoniously, under Mendelian rule, for 1500 million years. But at some point, parental care started to evolve. In the past 500 million years, the complexity of life has increased in many creatures – in flowering plants, in insects, and particularly in vertebrates – and the parental care of the offspring, after the egg is fertilized, has at least contributed to this rise in complexity. The mammalian maternal–fetal relationship is the most complex form of parental care, and it exists in the most complex kinds of life. The most intimate, and the most prolonged, forms of pregnancy are in the great apes, including human beings. There is an evolutionary tension between the intimacy of the physiological connexion, and the slight genetic difference between the mother and fetus. Natural selection has been able to favour subtly uncooperative genes. It has favoured some mild gynaecological dirty tricks and spin-doctored hormonal communication. Moreover, the genes of the mother and fetus are so intimately connected that they may have been able to find some partial ways round the Mendelian veil of ignorance. The evolutionary consequences can almost only be guessed at, but may extend to the wiring of our brains, the working of our minds, and the structure of our social relations. No general conflict-resolution procedure has evolved against the post-Mendelian lawbreaking genes. A conflict is, over evolutionary time, persistently present and may always sneak in and reduce the efficiency of reproductive or social arrangements, at least temporarily. The conflict between parental and offspring genes has certainly shaped, and may have limited, the evolution of complex life.

9
The human condition

Oh my soul, aspire not to immortal life
But exhaust the limits of the possible

Since the origin of life 4000 or more million years ago, or at least since the origin of the eukaryotic cell 2000 or more million years ago, the range of live complexity has been expanding and the upper peak elevating. The reason is probably the one we looked at in Chapter 2: there is often an opportunity to make a living by evolving to be more complex than any existing life form. I see no reason why these opportunities, which have been opening up from time to time throughout the history of life, should suddenly be exhausted with the evolution of human beings. I expect they will continue to open up in the future. Just as our ancestors found, in an evolutionary sense, a way of life that was more complex than the great apes of the time, so some future subpopulation of humans (or some other relatively complex species) could find a way of life that is more complex than ours. The extra genes of our more complicated descendants might be used for any number of purposes, depending on the challenges of current and future environments. I shall consider in the next chapter what a superhumanly complex life form might look like. If infectious disease, for instance, will make or break our future, we could use some extra genes for disease resistance. If it is computers and artificial intelligence, some extra genes could enable us to learn artificial intelligence in the way that we now learn our native languages. But in this chapter it does not matter exactly what the creatures would look like. We only need to assume that an ecological opportunity exists, or will come to exist, for some kind of life that is more complex than us.

The opportunity probably exists; but that does not mean it will be seized. The life form also has to be internally feasible. A more complex life form than us would use more genes than us, and experience more problems than us with copying error and subversively selfish genes. Human beings may already be at the upper limit of live complexity that natural selection can keep in order. An increase in complexity beyond us would then either be impossible, or have to wait for some new mechanism to prevent, correct or purge genetic error. At the origin of life – and immediately after it – the length of the replicating molecule was limited to a few hundred units. There was an opportunity for more complex life, but it did not evolve until enzymatic DNA (or RNA) copying had been invented. During the longer period of prokaryotic life, there was again an opening for more complex life, but it was not filled until after the eukaryotic cell had evolved and solved the problems of gene conflict. Now I want to ask whether human beings, with their few hundred mutations per breeding event, have hit the current limit of the biologically possible. Depending on the answer, the future evolution of complexity may either proceed by the same evolutionary processes as are operating now, or await the evolution of some new enabling mechanism.

Are human beings possible in theory?

The Chinese whispers game of Chapter 3 illustrates how a life form would be impossible if its mutation rate were so high that a parent made a copying mistake every time it produced an offspring. Natural selection would be unable to stop mutation from randomizing the life form. For the life form to persist, the average parent must produce at least one offspring without copying mistakes. Human beings appear to violate this criterion. We seem to make more than one damaging mistake in every offspring. Our existence might seem to be a paradox; but (as we saw in Chapter 5) a life form that uses sex can theoretically exist despite an error rate of more than one per offspring.

It is one thing to show that sexually reproducing life forms can exist with an error rate of more than 1; it is another to show that they can exist with as high an error rate as ours. Human beings make more mistakes when they copy their DNA than any other form of life on Earth. Every human being is conceived in 200-fold copying error. It is not known how many of our 200 mutations are harmful, but even rigorous accounting cannot squeeze it below about two, and a figure of five to ten may be more realistic. I should not be surprised if it is twenty. These big numbers – bigger than anything known in any other species – are a consequence of our complexity. Complex life forms contain many genes, and more genes means more mutations. We also live long, and have long generation times, which contribute to our complexity because we can acquire further information in our lifetimes by learning. Mutation rates go up as generation times lengthen, because the DNA is copied more times per generation if the generation time is long rather than short. Our long generation time probably also lies behind another source of error, which we met in Chapter 7. Chromosomal aberrations multiply over time in the cells that are destined to produce eggs. About 50 per cent of human conceptions have a botched number of chromosomes, whereas the percentage in rabbits or guinea pigs is negligible. These errors in the number of chromosomes are in addition to the 200 copying errors. It is as if, when you order a multivolume set of books, not only does every set have 200 printing errors, but half the time they send you a mismatched set of volumes.

Errors in chromosome numbers may be passive accidents, like copying errors; but they may also be caused by actively selfish genes. Male and female human beings may evolutionarily have travelled different roads as human life spans have stretched out – roads that correspond to the two main themes of this book. Men manufacture sperm throughout their lives, copying their genes many times. There have been about forty cell divisions in the reproductive cell line of a human male at the age of puberty, when he is about 13 years old. The sperm DNA is then copied every sixteen days, or twenty-three times per year. A 20-year-old man's sperm have been copied 200 times, a 30-year-old's 430 times, and a 40-year-old's over 600 times. Copying mistakes happen when DNA is copied, and human sperm contain more copying mistakes than those of most, if not all, other species. The sperm of an average adult male rat, for instance, contain DNA that has been copied only 58 times. Women do not make eggs throughout their lives. A woman has her full supply of eggs, with about 33 cell divisions behind them, when she is still a late-stage fetus. Her DNA is copied less, and probably contains fewer copying mistakes, than that of a man. But egg production is deadly for half a woman's genes, and her suspended meiosis may give selfish chromosomes the time to subvert the system. Errors in chromosome numbers accumulate over time, ultimately leading to menopause. When a 30-year-old woman breeds with a 30-year-old man, his DNA has been copied 430 times against her 33. There may be thirteen times as many errata in his scriptures, and about 185 of the 200 copying mistakes in each human conception may come from the sperm; but she is more likely to send an extra copy of the book of Leviticus.

Here I want to concentrate on the copying errors, rather than the errors in chromosome numbers. The fraction of fertilized eggs with chromosome errors is high in humans, and may be as high as 50 per cent; but that still leaves 50 per cent that are error-free and available to carry the human system forward. The 200 copying errors raise more of a question: it is not obvious that a system can exist with such a high error rate. The two kinds of error do interact, however. Sex can help with the copying mistakes, but it cannot help to purge the extra chromosomes. It takes as much death to clear chromosomal errors from a sexual life form as it does from a clonal life form. The copying errors that sex may help with have to be purged from the 50 per cent of fertilized eggs that at least manage to get their chromosome numbers right. Copying errors become more difficult to clear if the fraction of chromosomally abnormal eggs is higher. During the evolution of complexity, generation times have lengthened and gene numbers have increased. Eventually, the complexity of life hits the ceiling when the number of copying errors is just too many.

But how many is too many? What can the error rate go up to, if a sexually reproducing life form is to be indefinitely sustainable – 2, 20, 200, or what? The rocket scientists have written equations about this question, but the real answer remains unknown. What we can do is sketch out the main possibilities. At one extreme, sex can lift the roof off, enormously increasing the tolerable error rate of a life form. It could probably go up into the thousands, and certainly it could go higher than even the highest estimates for existing life on Earth. At the other theoretical extreme, sex makes no difference to the upper limit on the error rate. Sex helps to clear out copying errors only if a crucial condition is met, a condition that we came across in Chapter 5. Mutations have to do escalating damage as their numbers increase: two mutations together must hurt the organism more than expected from each by itself. The arguments I outlined in Chapter 5 suggest that the condition is met, but they are inconclusive. If living creatures do not show the 'escalating damage' relationship, then sexually reproducing creatures cannot exist with an error rate of more than 1, any more than clonally reproducing creatures can. Finally, sex could have an intermediate effect, revving up natural selection, but perhaps by only a factor of two or so. The tolerable error rate would then be 2, instead of 1. Whether the tolerable upper limit on the error rate is 1, a bit more than 1, or some huge number perhaps up in the thousands, depends on the details of how natural selection acts. We need more work on those details.

What implications does this have for human evolution? The most apocalyptic possibility is that we are in a mutational meltdown. If the upper limit on the error rate is 1, then our error rates of 2–20 would be too high for natural selection to cope with. We should be mutating ourselves to extinction. A more complacent conclusion could be drawn if sex has raised the upper limit on the error rate to a thousand or more. Selection would be comfortably destroying mutations as fast as they are created, and *Homo sapiens* would be evolutionarily motoring along with more important worries than mutational error. Or it could be that the error rate has an upper limit at about the level in humans now, at a bit more than 1 per offspring. Natural selection would then be struggling to keep up with genetic error. We should have a serious problem with genetic disease, but not so serious that it threatened us with extinction. If the last conclusion is right, human beings are the climax of complexity, at least with existing biological mechanisms. Complexity could not increase without some new mechanism to deal with error. If the complacent conclusion is correct, life can evolve to be more complex than anything that exists now, by the continuation of normal old evolutionary processes. If the apocalyptic conclusion is correct, complexity has gone over the top with us and would evolutionary shrink back over time, as the excessively complex species go extinct.

It might be thought that our existence alone is enough to rule out the most apocalyptic conclusion. If our mutation rates are too high for selection to cope with, we should not exist. But our existence in fact only suggests, and does not guarantee, that we are sustainable in theory. We may have been adding mutations at a higher rate than we have been eliminating them for some time. We should then be in a mutational meltdown, and destined for extinction by randomized DNA (if some other factor does not intervene before). The mutational meltdown operates on an evolutionary timescale, and a species could persist for thousands of generations, slowly randomizing its DNA, before it finally collapsed. Thousands of generations is tens of thousands of years, or even a few hundred thousand years for humans. How long have our mutation rates been as high as they are now? Our ancestors have probably made 200 copying mistakes per offspring, and put the wrong number of chromosomes in 50 per cent of their offspring, ever since our generation time evolved to the modern figure of 30 or so years. Experts disagree about when this happened. Some argue that chimpanzees and gorillas have similar generation times to us, which would push the origin of our long generation times back near the origin of great apes, at about 15 million years ago. Others would use a figure nearer 5 million years ago, when the human line branched off from the other great apes. Or maybe a figure of 2 million years would be better, if our modern generation length dates back to the origin of the genus *Homo*. Or it could be as recent as only a few hundred thousand years ago, when creatures indistinguishable from us – anatomically modern humans, as they are called – either had evolved or were soon about to evolve. The modern human mutation rate would look sustainable if it has lasted for 15 million years, but might be unsustainable if it has been around for only one or two hundred thousand years. There is some evidence to suggest that we are accumulating mutant genes at a higher rate than other species; but it does not convincingly show that we are in a mutational meltdown. I suspect that our mutation rates are older than us, and that chimps and gorillas have a similar mutation rate to us. In that case the apocalyptic conclusion is wrong. We have not, so far, been mutating our way to inevitable extinction.

La dolce vita

Our ancestors have coped, probably for 10 million years or more, with the highest mutation rate ever seen in evolutionary history. Somehow natural selection has acted powerfully enough to remove all the mutations as fast as they have arisen. But in more recent times, social change may have relaxed the force of natural selection, in some wealthy human societies, and allowed mutations to accumulate. For most of human evolution, natural selection has probably acted against mutations in humans much as it has in any other life form, if more extensively; but we, or some of us, may be moving into a new evolutionary phase.

I am going to compare two kinds of human society, which I shall call 'wealthy' (or 'rich') and 'traditional'. (The terms are imperfect, but then so are the alternatives.) The wealthy societies are the rich countries of America, East Asia, Australia and Europe; the 'traditional' societies are usually not whole nations, but societies within certain nations – societies of hunter-gatherers or simple farmers. There are other kinds of society in the world, such as the 'developing' nations, but I shall only mention them at the end of the argument, as a puzzling afterthought.

Natural selection may be relaxed in wealthy nations for two reasons. One is health-care. For example, during most of human evolution, since our brains completed their expansion around half a million years ago, babies who were born heavier or lighter than average had lower survival rates in the first month after birth. It is a classic example of natural selection, in which intermediate birth weights are favoured over either extreme. This pattern was shown for babies born in a London hospital in the 1930s, and in a New York hospital in the 1950s. But a series of measurements in Italy from the 1950s to the 1980s shows this force of selection being relaxed into near-invisibility. In Italy in the 1950s, the pattern was much like in London in the 1930s and New York in the 1950s – something similar to the human experience of half a million years. But by the mid-1980s, Italian babies of more or less than average weight had the same survival rate as average babies. The same trend has probably occurred in other rich countries too, and it is caused by improved obstetric medicine. In the 1930s, a gene making you a slightly lighter or heavier baby than average was a bad gene; now it is not. The genetics of birth weight is messy, and it is impossible to give examples of particular genes to illustrate the theory. But the same process is at work in other cases with simpler genetics. Phenylketonuria is a textbook example. Human beings who have this mutation cannot metabolize phenyl-alanine, a substance (an amino acid, in fact) present in a typical human diet; mental retardation and other horrors are the result. But the condition can be diagnosed, and people with the phenyl-ketonuria gene grow up more or less all right if they eat a special diet that lacks phenylalanine. I expect that the gene has increased in frequency in the societies in which it has been medically cured. In wealthy countries, one formerly harmful gene after another has probably increased in frequency as our medical understanding of them has improved.

Relaxed selection is not only caused by medicine. High living standards in general can have much the same effect. Medicine obviously keeps some people alive, but people are less likely to starve if food is abundant, or to poison themselves if food is fresh or refrigerated. If you have a slightly defective gene, such that you are slightly less efficient at digestion, and slightly less likely to survive a food shortage, or are slightly more sensitive to toxins in decaying food, then your genes will be more likely to survive in rich, than in poor, conditions. Similar arguments could be made for spacious, well-heated (or well-chilled) or well-ventilated housing. People can still enjoy life in wealthy societies even if they have some minor defects.

Demography is the other factor to consider. One rich country after another has undergone a demographic transition in the nineteenth and twentieth centuries, and the process will probably continue in the twenty-first century. Family sizes shrink and become more uniform. We can imagine an ideal, or a caricature, of middle-class society in which all the families are the same size – two parents and two children, perhaps – and everyone has the same low mortality. Natural selection could then have been brought to a stop. Mutations could accumulate in the population at the rate at which they occurred; the population could be in a pure mutational meltdown. The condition of a wealthy human population with relaxed selection is a mild form of the Mukai experiment that we met in Chapter 4 when measuring error rates. His experiment used soft living conditions and demographic tricks to stop natural selection from operating on a population of fruit flies. The result was that the DNA of the experimental fruit flies began to be randomized and the viability of the flies, when put back in normally rigorous conditions, declined.

Both medicine and demographic change have at most only relaxed the force of natural selection in us; they have not put a stop to it. Medicine is only imperfect, for reasons of scientific ignorance, limited availability and occasional bungling. The demographic transition has not gone all the way to the caricature in which every family has exactly two children. But it is not necessary for natural selection to stop completely for a population to tip into mutational meltdown. A biological species will normally have a rough balance between mutational input and natural selection to remove the mutations. Selection may need only to be a little relaxed for the rate at which mutations arise to exceed the rate at which they are purged. The meltdown may proceed slowly, but proceed it will.

Moreover, once the condition for the meltdown is in place, it is persistent rather than temporary or self-correcting. Mutations will continue to accumulate in the population, unless selection is restored to the same level as existed before it was relaxed – the level needed to balance the mutational input. If selection is relaxed we do not just accumulate some mutations for a while and then reach a new balance, with more bad genes per average person (and the mutations do not matter because medicine can take care of them). The mutations pile higher and deeper, until they bury us.

There are several uncertainties in the argument, and it also raises moral issues. But let us first look at the argument in a constructive and morally neutral light. The examples of birth weight and phenylketonuria illustrate the way that medicine relaxes the force of selection. But they do not show that the relaxation has genetic consequences. The argument would be more persuasive if we had direct evidence that bad genes are more frequent in rich countries. The best evidence is for colour blindness. The common kind of colour blindness, and the kind I am talking about here, is red–green colour blindness. People with red–green colour blindness are not colour-blind in the sense of seeing in black and white; they distinguish the red and green parts of the spectrum by light intensity rather than wavelength. It often makes no difference at all, and indeed many people who are red–green colour-blind are unaware of the fact. It is caused by mutations in a gene on the X chromosome, and is almost entirely found in men. Men have one X chromosome, and if it is defective they suffer from the defect. Women have two X chromosomes and can cope with a mutation in one of them (this is an example of how mutations can be masked if you have two copies of a gene – *see* Chapter 3, p. 72). Colour blindness is easy to test for, and we have heaps of evidence. A review paper written in 1962 described 7712 people from thirteen different 'traditional' societies, including hunter-gatherers or simple, non-industrial farmers, from Africa, Greenland and elsewhere, who had an average frequency of colour blindness in males of 2 per cent. In contrast, 436,853 people from ninety-nine samples of 'wealthy' societies had a frequency of about 5 per cent (and more than 10 per cent in some local samples). The 'wealthy' samples were mainly from European 'or ancestrally European' populations, but also from some East Asian societies; a high frequency of colour blindness is not simply a European trait. The frequency of the mutant genes that cause colour blindness appears, in rich nations in the mid-twentieth century, to have doubled from its ancestral level.

Myopia, or short-sightedness, is a second possible example of mutant genes that have increased in frequency because of relaxed selection. A much higher fraction of people are short-sighted in wealthy than in traditional societies. Charles Darwin himself had a mortifying lesson about the fact. As a 'hunting and shooting' type of young man he was proud of his eyesight. But then, in his early twenties, he sailed across the Atlantic in the company of some natives of Tierra del Fuego, who could see things miles off that were invisible to him and the British sailors. The Fuegians worked as lookouts on the boat, and would allude to their powers when some quarrel broke out ('me see ship, me no tell'). The high frequency of short-sight in rich countries is not in doubt, but the example is less convincing than colour blindness because the genetics is poorly understood. There is evidence for a strong genetic influence, but it comes from comparisons of twins, and studies of this kind are never convincing. Myopia may mainly have non-genetic causes, and its high frequency in rich countries may have nothing to do with relaxed selection. I suspect there is a genetic component, and mutation has been at work, but the example is nothing like as good as colour blindness, for which the actual gene has been identified.

Selection, therefore, may be relaxed in rich countries. We have a plausible argument, and evidence to illustrate it. Natural selection may have eliminated mutations as fast as they arose during the first few million years of human evolution, but now it may be struggling, or even failing, to keep up. What consequences would it have, for future human evolution, if mutations occur at a higher rate than they are eliminated in rich societies? The most dramatic consequences would follow if the habits now confined to rich people spread to the majority of human beings. Family sizes and mortalities may even out globally, and health-care may become less exclusive. Natural selection would then be relaxed in the whole species, and all *Homo sapiens* could conceivably be tipped into a mutational meltdown. We should go extinct because we are too complex: at a certain point in the evolution of complexity, life becomes clever enough to rebel against the cruel process – natural selection – that created it. Natural selection would then no longer protect this life form against mutational decay. Human civilization would be evolutionarily self-destructive.

Human extinction would be a disastrous consequence of relaxed selection, but there are several reasons to doubt whether it will happen. One is that natural selection might reimpose itself before our DNA had been randomized to the point of inviability. Another is that the relaxed selection regime of rich countries may not spread globally. The current differences between rich and poor countries may persist into the future. Then the medically cosseted rich (and their possibly accumulating bad genes) may be a near irrelevance for the human genetic future, because nearly all human reproduction would be in developing countries with high population growth. At this point the argument approaches terminal uncertainty. The problem is this. The case for relaxed selection compared wealthy with traditional societies. In both these kinds of society population sizes are fairly stable. Global population growth today, however, is dominated by a third type of society, with high population growth.* What I do not know is whether they have relaxed selection or not. They have not undergone the demographic transition, nor do they have as extensive health-care as rich countries. But there is another reason to suspect that selection is relaxed: namely, the high population growth. If you told an evolutionary biologist that you had two populations of a species, and one of them was growing rapidly while the other was stable or declining, and then asked which one had relatively relaxed selection, the answer would be that selection was relaxed in the rapidly growing population. The population growth itself suggests that natural selection is doing less killing. Mutations matter less

---

* The AIDS epidemic is a further complication; it is a decidedly unrelaxed selection pressure in some places.

there: you can survive and reproduce even if you do have an error in your genes. It would be in the constant population that selection was hitting harder. There it is more of a struggle to reproduce and insert your offspring into the next generation; selection will rigorously scrutinize genetic quality, and only the best will manage to reproduce. It is therefore possible that selection is relaxed in countries with high population growth for one reason, and in wealthy countries with stable populations for another. But I do not know that selection is relaxed in countries with high population growth: it seems plausible, but we lack evidence such as for the frequency of colour blindness. If selection is relaxed in countries with high population growth, then we have a stronger case that humans are close to a mutational meltdown; if it is not, then we do not.

Also, natural selection may not be generally relaxed in rich countries. Medicine has undoubtedly relaxed selection in some respects. Modern urban life, however, notoriously creates pressures of its own, and civilization may substitute one kind of selection for another rather than causing a net relaxation. Natural selection may, for instance, work by means of the stresses induced by social competition. We may be relaxed about mutations for poor eyesight, and we can load up on them. But they may be matched by a ferocious selective purge on some other genes. In modern life, natural selection may work against genes that cause your autonomic nervous system to freak out whenever a customer, or your boss, 'throws a wobbly', or you are late for the plane departure, or you motor round the bend and see – another traffic jam. But then you do not want to be born with a *mañana* gene, either – one of those genes that makes you insubordinate to the clock. *Mañana* genes and freak-out genes probably do not harm their carriers in traditional societies.

The condition for a mutational meltdown is that selection is relaxed in general, against almost all mutations. It is not enough for selection to be relaxed against a specific class of mutations. Colour blindness and myopia make an instructive comparison. Myopia is corrected by spectacles, and the spread of mutations that cause myopia is probably enhanced by health-care. But this relaxed selection is specific to genes influencing myopia: these mutations have ceased to be harmful in an environment that contains opticians. The quality of the population decreases a little, because short-sightedness is still an inconvenience, but relaxed selection against these particular mutations will not threaten us all with extinction. Colour blindness, by contrast, has not increased in frequency because of a medical cure. Its frequency has probably increased because of generally relaxed selection – the general ease of life – in rich countries: colour blindness was always a low risk to life, but became even less of a risk after the rise of civilization. It is this general relaxation of selection that may lead to a mutational meltdown. It allows slightly harmful, if not deadly, errors to accumulate in all genes. The colour blindness example (and the reasoning behind it) suggests that selection is indeed relaxed in rich countries. But it is only one example and it might be challenged. Colour blindness is not enough to prove that civilization has generally relaxed the power of natural selection, rather than substituting one force of selection for another.

I like to keep an open mind about relaxed selection. I suspect that selection is relaxed in wealthy countries, relative to hunter-gatherers and simple farmers with stable populations, and that errors are proliferating in civilized DNA. But I doubt whether the relaxation is enough to threaten our species, or some societies within our species. Even in wealthy societies, natural selection will work its way past the doctors and family planners, as we shall now see.

The four ways

The progress of civilization, or of wealth, has relaxed two of the ways in which natural selection acts against genetic error. Medicine, and the generally high quality of life in wealthy societies, means that people with mutant genes are less likely to die. Family sizes seem to be more uniform in rich than in traditional societies, and parents who contain mutations may have the same number of children as parents who do not. But even if these two factors were taken to the extreme, in which all families had two children and everyone had the same chance of survival, natural selection might still not have come to a stop. There are two other ways in which it can act against bad genes.

The first works in the early stages of the life cycle, long before birth. The life cycle, in which one generation breeds the next, begins with cells in the ovaries and testes that develop into eggs and sperm. These gametes then fertilize each other. The fertilized egg then develops as an embryo and fetus in the mother's womb for nine months, before a baby is born and grows up into an adult, who may reproduce in turn. Natural selection can act at any stage in this cycle to eliminate a mutant gene. Medicine, however, mainly works at the postnatal stage. Medicine – even surgical procedures – is used at the fetal stage, but the power of medicine generally reduces as we move back from birth to conception. As the power of medicine decreases, the power of natural selection necessarily increases; also, there is more scope for natural selection early in the life cycle because there are more potential victims at that stage.

The numbers of individuals increases as we move back through the cycle from adults, to children, to embryos. About four fertilized eggs are probably lost for every one that develops into a baby, but the numbers really take off at the gametic stage. The egg has a potential choice of sperm at the last moment of fertilization. The egg is surrounded by a gooey substance called the zona pellucida. Several sperm typically make it into this zone, but only one fuses with the egg membrane and moves on to final fertilization. And for every sperm that makes it into the zona pellucida, hundreds of millions are degraded in or flushed out of the female reproductive tubes. Hundreds of millions more do not even make it that far. The scope for selection among female gametes is less than for male gametes, but is still large. A human female contains a peak of about 5–7 million oocytes (cells at the stage just before the final meiotic cell divisions) but at most only a few hundred become eggs and at most only about ten become children. One in a million oocytes ever develops into a child; the other 999,999 are lost.

No doubt many of the genetic deaths at the early stages of the life cycle are non-selective, unrelated to the genetic make-up of the sperm, egg or early embryo that is lost. But that does not mean natural selection is impotent at these stages. Much of the death later in the life cycle – among children, or adults – is probably also non-selective. Natural selection would still do far more work earlier than later in the life cycle even if a much lower fraction of the deaths earlier on are selective.

A sceptic could argue that natural selection is unable to act in the early stages of the life cycle. The reason is that few genes are switched on at these stages. A gene has to be switched on and has to influence the cell, or body, it is in before natural selection can pick out the defective versions. Natural selection removes error like someone who assesses the accuracy of a piece of holy writ by watching the behaviour of its users, rather than by looking at the text itself. You can look for mistakes by going through the text, letter by letter, comparing it with the original. In the DNA, this is what the proofreading and repair enzymes do, before the genes are ever used to build a body. Alternatively, you can look at the effect of the text in the world. You do not read it; you just look at what effect it has on its users, and destroy the copies that produce defective results. That is what natural selection does. But the procedure only works for the parts of the manuscript that are expressed while you are watching. If the manuscript has one set of instructions for how to behave on Monday, and a second set for Tuesday, then you cannot detect errors in the instructions for Monday by watching what people do on Tuesday. If most of the DNA instructions in the egg are silent, you cannot detect mutations by looking at the egg. What we need to know is how many genes work in the early stages of the life cycle.

We know something about how many genes are expressed in the embryo. Almost 60 per cent of the genes in the DNA are probably expressed at some stage in the embryo and fetus. This will include all the genes that code for basic cellular housekeeping. It also includes the genes for the general working and coordination of the body, and the genes that guide development. A fetus with a mutation in any one of these genes may be more likely to die, and the many conceptions – perhaps as many as 80 per cent of them – that are lost before birth probably partly reflect the workings of natural selection. The fraction of genes that are expressed in the embryo is the upper limit of the fraction that can be scrutinized by natural selection. Many of the genes that control eyesight, for example, must be expressed in the embryo because the eyes have been built by the time of birth. However, I guess that an embryo with genes for defective eyesight is no more likely to be lost than one with genes for perfect eyesight. The genes cannot be tested until after birth, even though they are expressed before. It is interesting that the examples of relaxed selection that we looked at above were for genes that influence specifically postnatal skills. In all, selection probably works on many, but not all, of the genes in the DNA in the early stages of the life cycle.

Our scientific knowledge fades out as we move back to sperm

and eggs. We have some evidence that mutations are weeded out in the cells that develop into eggs: mutations are found in more of the early-stage cells than in the final fertilized eggs. How the ovaries detect the mutant cells is unknown. We also have evidence that mutant sperm are less successful, and we also do not know how they are selected against. Suspect number one consists of certain design features in the female reproductive system. This system is notoriously hostile to sperm, springing a series of tricks on its long-suffering visitors – tripwires, booby-traps, buckets of water balanced on the door-tops – all of obscure purpose but the subject of devoted biological research. It is possible that they test the genetic quality of the sperm. It is also possible that they do not.

Defective sperm and eggs in the testes and ovaries could be removed even without checking the DNA directly and even if no genes are expressed. Let us return to the analogy with a piece of holy writ. You are looking after your soul and you go to the local monastic manuscript store to buy some holy writ. The storekeeper offers you a choice of copies, but you happen to know that some were copied in the normal calm of the cloister and another was copied while the scribe could hear a pair of stiletto heels scrunch on the pavement outside. You may be able to guess which this copy was even without trying it out on yourself, or comparing it with the master-copy: it may have the date written on it, and you know when the nun was visiting; or it may have tell-tale doodles in the margin. Would you trust your immortal soul to this copy? I think you would buy one of the others. My point is that the ovaries may be able to tell which oocytes have experienced distracting or stressful conditions. There are so many cells to play with that the 'nurse' cells would only need the slightest suspicion about an oocyte for it to make sense to smother it. Selection of a sort has then acted against some female proto-gametes that are more likely than average to contain copying mistakes, and the eggs that finally float down the fallopian tubes will contain more accurate DNA than the average for the original multimillion-cell pool.

Medical advances have done little against natural selection in the early stages of the life cycle. New reproductive technologies may do more in the future, but I am concerned in this chapter with existing, widespread medical practices only; the next chapter will speculate about future medical technologies. I should just remark that although some fertility treatments can allow defective early-stage cells to reproduce, the usual practice is to favour 'good' embryos. Future reproductive technologies may be more likely to assist, than to relax, the forces of natural selection against genetic error. But whatever the future brings, civilization so far has hardly relaxed the force of natural selection early in the life cycle.

The fourth way in which selection probably acts against bad human genes is by whether an individual manages to marry and produce a family. ('Marry' here must be understood in an extended, perhaps old-fashioned sense. It refers to pairing, with or without the ritual of marriage; and it implies breeding, as if people who marry breed and people who do not marry do not breed. 'Married' here means 'paired-off breeder'.) This is the kind of sexual selection we looked at in Chapter 5. In a sexually reproducing species, selection against bad genes can be improved if individuals of high genetic quality are preferred as mates. In the frogs we looked at in that chapter, there was nothing like human pair bonding; the frogs were polygamous. Male frogs with bad genes were less likely to breed, because female frogs did not like the noises they made. A similar process may well operate in us, but probably in both sexes because we are mainly monogamists. In humans, males choose among females as well as females among males, and females compete for valuable males just as males compete for valuable females. A bad gene in a female body may lose out in the marriage market as well as, if to lesser extent than, a bad gene in a male body.

It is not simple extrapolation that leads me to suggest that genetic quality influences human marital decisions. There is some suggestive evidence. It has been known about for ages; Darwin discussed it in 1871, and indeed I am taking the argument from him. The evidence comes from the death rates of single and married people. The pattern was first demonstrated in the mid-nineteenth century, and Darwin quoted a huge French study and a Scottish study, both of which showed that single men had about twice the death rate of married men, between the ages of 20 and 30 years. The pattern has since been shown in every country in which it has been looked for, and applies to all ages and both sexes. A comprehensive review of the evidence up to the 1960s found death rates of single men were about 1.8 times higher than those of equivalent married men, and death rates of single women were about 1.5 times higher than those of equivalent married women. Much the same effect was found in America, Europe and Asia. The effect may be different in humans now because marital and reproductive patterns have changed; but I have not found any more recent evidence. In the evidence we have, being single is one of the biggest risk factors in human life. If you look at a list of the big risk factors for human beings, the top ones in the list are being old, being poor, being male and being single. Being single is about as bad as being male, or poor; it is actuarially equivalent to smoking about a pack of cigarettes per day. What is going on here?

Two things are probably going on. One is that marriage itself probably improves your health. You may have two people looking after you rather than one; you do not have to perform dangerous pre-mating displays; you may enjoy economic and other forms of synergism; your status may go up. That said, you do have two people to look after, and married people are the people who breed. In humans, as in other species, breeding increases your death rate. The more children you have, the earlier you die. Single people have 1.5–2 times the death rate of married people, despite the fact that reproduction will have increased the death rate in married people. In all, it is not certain that marriage will improve your health, but on balance it probably does.

The other factor is that people with poor health are discriminated against in the marriage market, and the class of single people are the unhealthy left-overs. (There is a nationalist joke, which is told in many neighbouring countries around the world. I first heard it from a Scot, in England: 'whenever a Scot crosses the border into England, the average intelligence of both countries goes up.' It looks as if, whenever a single person marries, the average health of both the single and the married class goes down.) The factors that people use to choose their sexual partners are a popular research topic, and one of the factors that has been emphasized is bodily symmetry. People with more symmetric bodies and faces – that is, their left side is a good mirror image of their right side – are seen as more beautiful and are sexually more successful. Body symmetry is also associated with good health. A preference for beauty could then be part of the mechanism by which healthy people succeed in the marriage market.

We have two explanations for the higher death rate of single people: marriage improves health, and health improves marriageability. The experts have made arguments both for and against each explanation, but the attempts to eliminate either of the factors have all proved inconclusive. Both factors are plausible enough, and they both help to explain the full range of observations. I expect they are both at work. However, only one of them matters for the main argument here: the marital preference for healthy partners. We also need another argument, that there is a genetic component to health. Some people will be unhealthy for accidental, environmental reasons – they caught some disease in childhood and never fully recovered: when members of the opposite sex pass them by, no bad genes are being selected against. But other people will be unhealthy for partly genetic reasons, and their life choices or failures in the marriage market are a form of selection against bad genes. It is notoriously difficult to prove genetic effects in humans, but it is very likely that there is some genetic influence on health: genes are known to influence many things; health is a fairly crude variable, which will be influenced by many factors; and we are not exactly short of evidence of genetic influences on disease. In all, therefore, I expect that marital discrimination against unhealthy partners will in part be selection against bad genes. Mate choice in humans then discriminates against bad genes, just as it does in frogs, butterflies, stalk-eyed flies, and every other species that has been appropriately studied.

Medicine, and its ability to keep people alive who would otherwise be dead, is an amazing achievement of human civilization. It naturally dominates discussion of natural selection against human bad genes. But selection at the postnatal stage, among children and among adults, probably only ever accounted for a small fraction of the mutation purge in human beings. The amount of postnatal death is low compared with the amount of prenatal death, and most genetic defects probably show up at early stages, by reducing the quality of the embryo. The demographic transition is also one of the great events of modern history, but its Darwinian consequences can be deceptive. Family sizes may shrink and become more equal, but we also need to think about the people who do not pair off at all. Natural selection may not be able to work much, in a modern middle-class society, by differences in family size; but it may be as powerful as ever when we look at who succeeds in breeding and who does not. The size of the unpaired, non-reproductive class has expanded in the past two centuries, and is expanding. We are probably – partly, indirectly, unconsciously – influenced by gene quality when we choose someone to breed with. Although the demographic transition has reduced one component of selection, it has not eliminated, and may have intensified, another. Life is still difficult for most harmful mutations in human DNA. They get killed in gametes, they get killed in embryos, they get snubbed in the marriage market. Civilization has relaxed some kinds of selection, and may be allowing some of our genes to decay. But my guess is that natural selection still acts against most bad genes with much the same force as it ever did. Where wealth accumulates, most genes don't decay.

The morals of natural selection

What conflict, if any, is there between our morals and the future evolutionary fate of the human species? The answer matters both in itself and because it may help us to predict whether the current medical and demographic trends will persist. Medical and demographic change has only partly reduced the power of natural selection. If morality is on the side of these changes, they are more likely to persist: at least, some people will try to make them persist. If it is not, then this is one less reason to expect them to persist: at least, some people will not want them to. Here I shall argue that natural selection is morally objectionable only in some of its actions, and bad genes can be purged without causing any moral offence.

I distinguished four ways that natural selection may eliminate bad genes in humans: by differences in (i) survival late in the life cycle, (ii) family size, (iii) survival early in the life cycle, and (iv) chance of marriage. The main moral case against natural selection is for factor (i). The progress of civilization largely consists, in a Darwinian sense, of putting a stop to this kind of natural selection. It is a seriously nasty and morally wicked, process. It took a macabre analogy with the game of Chinese whispers to explain natural selection; and real natural selection is morally rather like introducing tigers into a children's playground. Natural selection of type (i) works by death. I do not want to be a victim of natural selection, and I do not want any of my friends and relations to be victims of natural selection. I do not really want anyone to be a victim of natural selection. It is generally accepted that human morality is in conflict with the Darwinian process. Thomas Henry Huxley famously expressed it in a lecture on 'Evolution and Ethics' in 1893. In his words, 'that which is ethically best – what we call goodness or virtue – involves a course of conduct which, in all respects, is opposed to that which leads to success in the cosmic struggle for existence.' We are therefore morally justified when we prevent the action of natural selection, and if our success creates a problem of genetic decay this is not a reason to go back to natural selection to solve it. The solution would be at least as bad as the problem. As Darwin said, 'if we were intentionally to neglect the weak and helpless, it could only be for a contingent benefit, with an overwhelming present evil.' The 'overwhelming present evil' is the real suffering of people exposed to natural selection. I am less certain of what he meant by the 'contingent benefit', but for our purposes it can mean that humans are not definitely heading into a mutational meltdown, and restoring the full rigours of natural selection may not be needed to save us from anything.

The first two kinds of natural selection are morally different. It represents clear moral progress to prevent premature death in adults and children. In so far as civilization does put a stop to this kind of selection, civilization is morally progressive. But the same can hardly be said for differences in family size. Family sizes do seem to be more equal in wealthy middle-class societies than in agricultural, early industrial and hunter-gatherer societies; but this is something that, from a moral viewpoint, has just 'dropped out' of social change. It is hardly a moral imperative, like health-care and medicine. Moral factors may have been at work. Family size, for instance, may have become more equal because wealth is more equally distributed. The process could continue into the future. This argument, however, is much less certain than the one for medicine. A successful medical act undoubtedly counteracts natural selection, and is undoubtedly morally good. How much wealth has to do with the demographic transition, however, and how much morality has to do with wealth, are both open to doubt. It is unambiguously morally offensive if someone is left to suffer and die for want of medicine. It is at worst less morally offensive, and at best morally neutral, if some parents choose to have three children and others choose to have one ('at worst' and 'at best' refer to the reasons for the parents' choices – whether 'at worst' they were constrained by inequalities of wealth, or 'at best' they acted on morally neutral individual preferences). It need not be one of the aims of civilization to equalize family sizes.

Natural selection at the earliest stages of the life cycle is probably morally neutral. Some gametes succeed, others fail: I doubt many people mind about the gametes that fail to become full human beings. As we move from the gametic stage forward through the life cycle, the moral situation changes. At some stage in pregnancy, a prenatal death is morally much like a postnatal death and we actively use medicine to prevent it. For natural selection late in the life cycle, the same moral argument applies as we looked at above: indeed, these deaths are 'late' in the life cycle for the purposes of the argument here. I have in mind genetic 'deaths' before fertilization, or in the day or two after fertilization, when I suggest that early natural selection is morally neutral. There is a moral incentive to apply medicine earlier in the life of the fetus than may be possible now; but I doubt there will be moral pressure to prevent genetic deaths among gametes.

What about the fourth factor, the chance of marriage? I do not think anyone would regard it as moral progress to put a stop to this kind of selection. To stop it, we should have to equalize everyone's chance of marriage, such that every person had an equal chance of reproducing – or trying to reproduce. This would mean something like drawing lots to determine whether you went into the single or married class. It would hardly be moral progress to replace individual choice with a marital lottery; indeed, it would be the opposite of moral progress. In so far as the human marriage market does select against bad genes, this kind of selection will persist into the future.

The future is unlikely to bring in marriage by lottery, but it might bring other changes that could influence the way that marriage discriminates against bad genes. Suppose, for instance, that the quality of our genes is displayed in the beauty, in youth, of our bodies and personalities, and that marital choices are made according to beauty. If future cosmetic surgery manages to give us ideal, and identical, looks, and cosmetic pharmacology manages to give us ideal, and identical, personalities, then this last hold-out of natural selection will be lost and marriage will become a genetic lottery. Analogous futurist arguments may be imagined for other factors that may influence marriage, such as wealth and power. It seems more likely, however, that the marriage market is inherently competitive, and our marital criteria will move ahead of medical and social change. I do not know whether these criteria will always discriminate against bad genes, but it would not surprise me. We can cross that bridge when we come to it.

The moral case against natural selection – real as it is – is only strong against one of the ways natural selection acts. Morally we should aim to provide medical protection after all but the earliest stages of the life cycle; but differences in family sizes, at least in wealthy societies, and the chance of having a family at all, are much less morally offensive. Moral progress does not require us to put a stop to natural selection, and natural selection in future civilizations (in so far as they are guided by our moral values) will probably continue to eliminate many of the copying errors in our DNA.

In this chapter, we have been concerned with whether human beings are at, or even beyond, the limit of Earthly complexity. Our mutation rates are certainly high, and there are hints that we are starting to stretch the limits of the possible. For instance, if you put a fertile buck and doe rabbit together, you can make new rabbits without difficulty. But it takes a fertile human male and female an average of five oestrus cycles of effort to produce a baby, and the main reason why four of these are unsuccessful may be genetic error, in the form both of rogue chromosomes and copying accidents. Natural selection at the earliest life-cycle stages acts to remove these errors and our relative infertility, as compared with other species, is the consequence. A life form that lived much longer or used many more genes than us might approach sterility. But there is no strong theoretical reason to conclude that we are near the limit of live complexity. The key question is how effectively sex helps natural selection to clear error. It is theoretically possible that we are near the limit, but it is also theoretically possible that there is a huge evolutionary upside that is open above and beyond us. The evolutionary mechanisms that are already operating would then bring more complex beings than us into existence in the future.

We also considered whether civilization, in the form of medical and demographic trends, might relax the force of selection and limit the evolution of complexity. It looks to me as if civilization relaxes selection against only some kinds of error, rather than generally relaxing selection against all error. Civilization will not end in a mutational meltdown. And in the future we may invent error-correcting as well as selection-relaxing biomedical technologies. The question will then be how civilization will contribute to, rather than limit, the evolution of complexity.

10
A complex future

Reprogenetics 2000

The year 2000 may prove to be a good approximate date for the revolution in human genetic and reproductive technology. Test-tube babies are already a quarter of a century old, and it will be another quarter of a century before some of today's frontier technologies are established. Genetic technology, in particular, is unlikely to take off until well after the initial stage of the human genome project – which will provide a complete transcript of human DNA – is complete in 2001. But a quarter of a century here or there will look like a minor matter in a few centuries' time. New reproductive and genetic technologies are appearing at a spectacular rate now, by any standard. The evolutionary themes of this book have implications for several of these technologies, and vice versa: the technologies have implications for evolution. This chapter is about these implications; the discussion is futuristic, and therefore conditional and uncertain.

Life deals with error by preventing it from happening, by correcting and concealing it and by purging it. The use of polymerase enzymes and DNA, for instance, helps to prevent errors from happening to begin with. Another set of enzymes corrects most of the errors that nevertheless arise. Proofreading and repair enzymes correct errors in the code; developmental troubleshooting enzymes correct the expression of faulty codes without correcting the code itself. Finally, sexual reproduction may have evolved to shuffle the errors that slip through all these enzymatic defences, such that they can be purged efficiently, in multiple form. Mate choice in favour of error-free members of the opposite sex also increases the efficiency of the purge. An analogous set of stories could be told for actively harmful selfish genes. Errors are prevented; errors are corrected; errors are efficiently purged. The evolutionary significance of any future genetic or reproductive technology depends on which of these categories it falls into. We saw in the previous chapter that new anti-error mechanisms may or may not be needed before the complexity of life can increase beyond us. If they are, then technologies that are now at the stage of research and development could ultimately launch us into the evolutionary future. And even if they are not, the technologies can still have evolutionary consequences, as we shall see.

Dolly folly

The first possible future technology I wish to look at is cloning. I am concerned with full reproductive cloning here, in which an offspring would be made from an exact copy of the DNA of one individual parent. The technical difficulties of cloning a human being no longer look insurmountable since the birth of a lamb named Dolly in 1996. (The story broke in February 1997, when Dolly was six months old.) I expect that cloning will be used by a minority of human beings who for medical reasons have no other reproductive options. But could human cloning ever become widespread: could most, or even all, human reproduction become clonal?

At this stage, the Darwinian answer has to be: probably not. We need sex. We may need it to clear our harmful mutations. A sub-branch of human beings who went in for clonal reproduction would also be signing their progeny up for a mutational meltdown. They would undergo rapid genetic decay, as mutations accumulated faster than they could be eliminated. I do not know how many generations it would be before every offspring was so loaded with genetic defects that it would be dead; the details would depend on the exact cloning procedure, but cloning could not last long on an evolutionary timescale. Any one individual might be successfully cloned: the offspring might have ten to twenty extra bad genes, but survive them. But the process is unsustainable, and cloning could be at most only an occasional, minority habit. The scaremongers like to imagine a mad dictator who will try to clone himself a thousand-fold. My forecast is that the clone would be sick, and destined to collapse under the burden of its own copying errors.

The same problem can be seen in more individual terms. You are contemplating reproduction, and deciding between clonal and sexual reproduction. On average you make, say, ten to twenty harmful mistakes when you copy your DNA. If you reproduce sexually, the mistakes will be cleared and your offspring will be genetically as good as you are. But if you reproduce clonally, most of the mistakes will not be cleared and your clone will have to live with these defects because of your decision. It would be analogous to breeding sexually, and then inflicting ten to twenty minor injuries on your newborn child. If the harmful mutation rate is less than 10–20 per offspring then the damage would be less, but the morality would still be indefensible.

Moreover, the problem with cloning does not depend on Kondrashov's theory of sex in particular. We do not know for sure that sex exists to purge bad genes. Another plausible theory is that we need sex to keep up, evolutionarily, with parasites and disease. In this case, the clones would succumb not to bad genes but to disease. Cloning yourself would be like breeding sexually and taking your children into a plague-stricken city, where their chance of dying from disease is doubled. The anti-disease theory of sex may turn out to be wrong too, but it is highly likely that sex has some advantage, and that the advantage is big. Sex would not have evolved, and been retained, unless it had some advantage. We saw in Chapter 5 that a sexual form of life will reproduce at only half the rate of an equivalent clonal form. The halved reproductive rate of sexual forms is probably made up for by a difference in quality: the average sexual offspring is probably twice as good as an equivalent cloned offspring. We can expect that a cloned human (or sheep) will on average have half or less the quality of a sexually reproduced offspring. A halving in quality is serious: being cloned is probably analogous to losing an arm or a leg.

We should always be careful with new technologies and new interventions. In the case of cloning, we should be tampering with – indeed, cutting out – a feature (sexual reproduction) of ourselves that may have a two-fold advantage, though no one can be sure what the advantage is. Cloning could be, in a delayed-action way, rather like volunteering for the surgical excision of your heart during the pre-Harvey era when the function of the heart was unknown. The surgical technology may be space-age, using the best composite material knives, but the basic problem lies in messing with a design feature of our bodies when we do not understand the design principles. This point applies even though we have reason only to think, rather than to be certain, that sex is advantageous. Sex may exist for some reason that no longer matters to us, and we should be unharmed if we dispose of it. It would be a good idea to find out first. Only when we know why sex exists will we be able to say whether cloning is safe or not, or what kinds of cloning are safe. Meanwhile, it makes sense to assume that sex does exist for some good reason, and that cloning will contribute little to the human evolutionary future.

> **Holding hands at midnight,**
> **'Neath a starry sky,**
> **Nice work if you can get it,**
> **And you can get it if you try.**

Gene therapy may be another story. Gene therapy means medically curing defective genes, either by replacing the bad genes with good ones or neutralizing the bad genes, or by some other way-out future-tech. The subject is indeed futuristic, but the gene therapeutic techniques have already been made to work. Genes have been replaced in mice using techniques that would probably also work with humans – at least there is no reason to think they would not. There is also a simpler method, called 'anti-gene therapy', in which bad genes are switched off rather than replaced. It can help, for instance, when we have two copies of a gene, one bad and the other good, and the bad copy is causing damage in the body. If the bad gene is silenced by anti-gene therapy, only the good version of the gene is left working. The techniques as yet exist only as crude prototypes, but they are enough to show that the topic is realistic, not starry-eyed futurism.

This is not to deny the large gap between inventing a technology and applying it powerfully. Even the limited technology that does exist would have to be shown to be safe in humans, and in any case we could not do much with gene therapy now, relative to its potential, because few defective genes have been identified. Geneticists have, in absolute terms, identified a large number – maybe in the hundreds, maybe in the few thousands – of defective genes, but they have only scratched the surface. A human contains 60,000 or so genes and every one of them will have many defective versions. Genetic defects can also exist in the bits of DNA that do not code for genes. When gene therapy does start to be used, someone's quality of life will go up every time a genetic defect is cured. The technology should not be sneered at even if it is only used in a minor way. But if gene therapy could be used to cure a large fraction of human genetic defects, it could become influential on the grandest, evolutionary scale. It would need some general method of recognizing defects, and we do not have that now, but I can imagine that the human genome project will enable us, in the foreseeable future, to identify a large fraction of human genetic defects. Gene therapy could then become almost a cultural version of a repair enzyme, analogous to the biochemical repair enzymes we met in Chapter 4. We saw in that chapter how the copying error rate of DNA decreased from about 1 in 10,000 to 1 in 10,000 million letters as repair enzymes evolved in early microbial life forms, and that the copying accuracy has arguably improved little or not at all since then. If we do manage to use gene therapy to drive down the error rate in our DNA, it would then be the first big event in DNA replication techniques for 2000–3000 million years.

Gene therapy works like a repair enzyme; other technologies may reduce the underlying error rate, before repair. Consider the idea of gamete freezing, or gamete preservation by some method other than freezing. In women, the genetic quality of the cells that become eggs tends to decay with time, perhaps because of genetic lawlessness (as we saw in Chapter 7). The chance that a 20-year-old mother will have a baby with an extra chromosome is negligible; the chance in a 40-year-old mother is several per cent; and soon after that the biological clock reaches midnight, with menopause. In the future we may be able to hold the clock hands still. A young woman could remove some reproductive cells, softly embalm them, and later revive them for a pregnancy at a time of her own choosing. One consequence could be a reduction in the mutation rate. Indeed the mutation rate could be minimized by taking the cells as early as possible, for instance at birth, though there could be a trade-off here with the need for informed consent. I think it is a realistic speculation that young women will in the future preserve some reproductive cells for later possible use.

Men could also freeze their gametes. An old man's sperm contain more mutations than did the sperm of that man when he was young. The consequences of this, if any, are uncertain, but the geneticist James Crow has joked that the greatest threat to the human genetic future is fertile old men. The incentive for men to preserve some of their youthful sperm may not be as high as the menopause-inspired incentive for women to preserve their youthful eggs, but the incentive is still there. The human mutation rate will decrease if either men, or women, or both, preserve their youthful gametes. Gene therapy and gamete preservation probably will not be the only future technologies that influence our genetic quality. But they do illustrate the two main categories that these technologies are likely to fall into. Future technologies can either correct error, like gene therapy, or prevent error from happening, like gamete preservation.

I am going on to consider the long-term evolutionary consequences of these new technologies, as they reduce the error rates in our genes. In order for them to have any consequences at all, they will have to make it to the market and be used. Some of the technologies are ethically controversial and if they are banned or unused then all bets are off. My guess is that they will be used. One reason is that ethically the most controversial idea is gene 'enhancement', in which an individual's genes are replaced with the aim of improving athletic, mental, or other abilities. Gene enhancement is irrelevant here. I am concerned with the use of technologies to reduce harmful error rates – with the technological analogues of error correction and improved copying accuracy. The harmful error rate will be reduced if people use new technologies to cure, or prevent, genetic disease. This use is ethically less controversial than gene enhancement: in fact it is ethically similar to normal old medicine.

People already face similar decisions and often choose in a way that suggests that people in the future will make use of new gene-curing technologies. For instance, people who carry genes that can cause disease are in some cases offered counselling about marital decisions. Some Jewish populations use a system in which potential marriage partners can (if they wish) find out whether they would be likely to produce children with genetic diseases such as Tay–Sachs disease; an anonymous, numbered record is kept of the genetic test results of each individual and the results can be made available (with consent) to prospective marital partners. Pregnant women who are genetically at risk of having babies with Down's syndrome or other diseases are in some countries offered a test, and if the test proves positive the woman is offered an abortion. What happens is that many, though not all, people choose to abort fetuses with serious genetic diseases, and not to marry (or not to breed) if it would tend to result in genetically diseased children. The decisions in these cases can be completely ghastly. It therefore seems likely that people will choose to make use of new technologies for curing genetic error, where the decisions will usually be much easier. The new gene and reproductive technologies could then influence our future evolution. What might that influence be?

An evolutionary watershed?

We saw in Chapter 9 that it is uncertain whether the complexity of life on Earth is limited, with existing biological mechanisms, to something about as complex as a human being. It is possible that far more complex life forms could evolve. Or we could be near the limit, or even over the limit. The answer depends on how well sex helps to clear mutational error, and that is a topic of current research. If sex is a powerful force to purge error, life can evolve to be more complex by existing biological processes. Then the more complex life forms of the future may be descended from any species now, though they are more likely to be descended from species that are already fairly complex, for reasons we looked at in Chapter 2. Gene therapy, gamete freezing and so on could transform the lives of individual human beings, but in the grand sweep of evolution they would be trivial. At the other extreme, if sex is a weak force, the new gene and reproductive technology could be the most momentous event in the history of live complexity for 2000 million years. It might deserve to be categorized with the handful of breakthroughs that have formed the narrative of this book: the origin of reliably replicating molecules, repair enzymes, the machinery of Mendelian justice, sex and gender. These mechanisms have enabled life forms with more genes and greater complexity to evolve, but they may have been exhausted by the time we reach human beings. Evolution would then be 'waiting' (in a sense) for some new device to enable a life form with more genes than ourselves to evolve. The new gene technology may possibly prove to be the device. The future of complexity on Earth would then lie with our descendants, rather than any other species, because we alone have the technology.

Even if the new gene technologies are not needed for future increases in complexity, they could still influence the form of future, post-human complex life. For example, a few pages back I poured buckets of cold water on the idea of widespread cloning of human beings. Another way of expressing my point is as follows. In evolution, increasingly complex life forms become dependent on each new error-purging, or error-reducing, mechanism as it arises. The mechanism becomes locked in. When sex evolved, it enabled more complex life to evolve. However, these life forms had so many genes that they would collapse if they abandoned sex. Such is the reason to doubt we could go in for cloning in a big way. We can no more do without sex than we can do without our proofreading enzymes. I stand by this argument, on its own terms. But suppose now that we acquire some technological error-reduction mechanism. Evolutionarily, the mechanism can be exploited in two ways. It could allow, after millions of years of evolution, a future species to accumulate more genes and become more complex than us. I shall discuss this possibility in a minute. Another possibility is that we could use the new error-reduction mechanism to economize on our existing mechanisms. If the new gene and reproductive technologies become cheap and widely available, they could remove the problems I identified earlier, enabling us to clone ourselves if we wish. Indeed, they could even set up a Darwinian selection pressure for us to evolve virgin birth. We could evolve cloning biologically rather than technologically, though it might take a few million years by biological evolution. Sex is the messiest and most expensive of our error-fighting mechanisms, and our perpetually virgin descendants may multiply themselves with the error-permissive procedure of cloning balanced against the error-reducing procedures of gamete freezing and gene therapy.

Gene therapy could only enable us to clone ourselves if the reason why we have sex is to purge mutational error; the argument assumes that Kondrashov is right about sex. If sex exists instead to help against parasitic diseases, gene therapy is probably an irrelevance to the fate of cloned humans. (We might use gene therapy to insert disease-resistance genes, but this would be a case of gene enhancement rather than the error-correcting kind of gene therapy that I am concerned with.) In any theory of sex other than Kondrashov's, cloned humans are likely to have some huge disadvantage, whether or not gene therapy is available. My general conclusion still stands: we need to understand why sex exists before we go in for cloning.

We therefore have a two-pronged argument that life can evolve to be more complex than it is now. Either evolution can continue as normal, because sex allows a huge increase in the number of copying errors. Or evolution can be propelled forwards by the new technologies of error prevention and error therapy.

More of the same

It is a reasonable (if uncertain) prediction that more complex life will evolve on Earth in the future, or may have already evolved elsewhere in the Universe, but it is as much science fiction as science to speculate about particular life forms. I am now going to discuss some future, or other-worldly, life forms, but I do not think of them as real predictions. Indeed, if they were predictions, the one sure thing we can say is that they would be false. The future will be influenced by many factors: I should overlook some of them, and guess wrong about the operation of the others. But this does not mean that the exercise has no interest. It is a way of exploring the themes of this book, and broadening our view of life, particularly our own lives, on Earth.

What might a life form that is more complex than us look like? A hundred thousand genes of DNA code gives you a human being, but what would two hundred thousand, or two million, genes give you? We can begin by extrapolating from trends observable in wealthy human societies today. I do not mean that only the currently wealthy societies are on the road to the human evolutionary future, but it is one possibility to think about. One obvious trend is the time we take to insert ourselves in the information society. Education takes time, climbing the career ladder takes more time. In early modern Europe, you could graduate into the most advanced information-society career before you were twenty years old. Today in the USA it is not remarkable for people to be thirty years old by the time they are through law, medical or business school. The trend has not run its course; there is plenty of pressure to extend it. There is a steady supply of new knowledge, concepts and reasoning devices, and a steady pressure to squeeze them into our educational courses. In my subject, biology, most people think they know something that ought to be added to medical education, and are astonished at the lazy and hidebound medical establishment who refuse to add it in – to the great danger of future humanity. As if medical students were already short of things to learn! I hardly need to labour the point. Education and training could easily gobble up another ten years of our lives; we should just have to live rather longer to make it possible.

The existing demands of an information-society career use up a large fraction of the human lifespan. An additional ten years of education would be almost impossible. Almost every feature of our bodies has built-in obsolescence, and we could not increase our lifespans by tinkering with one or two genes. But the ageing of our reproductive cells has probably set the general rate of ageing of our bodies. Our bodies will have evolved to decay at the same rate as our gametes do, because there is no point in having a high-quality body after your gametes have all decayed by copying errors and chromosomal disruption. The chromosomes in an egg can be held in order for twenty or thirty years, but rogue chromosomes increasingly break loose after that. Almost every fifty-year-old egg is probably spoiled by disorderly chromosomes.

A very different life form from us – perhaps one that had evolved on another planet – could live longer if it lacked our eukaryotic reproductive constraints. The idea depends on two suggestions from earlier in the book; both were uncertain. In Chapter 7, we saw that the decline in egg quality with age may partly follow from the fact that only one of the four cells produced by meiosis becomes an egg; the other three are destroyed. The reason only one of the four becomes an egg may be, as we saw in Chapter 6, that mitochondrial forces have to be massed, and this can be achieved by unequal cell divisions in which all the mitochondria stay in one of the cells. If these ideas are correct, a life form that happened not to have mitochondrial DNA would not have evolved to destroy three of the four cells in meiosis. All four of the cells could produce gametes in both parents, just as males do in Earthly eukaryotic life. Then there would no longer be an advantage to selfish chromosomes that could position themselves in the lucky one out of the four cells; all four cells are equally good and the chromosomes should be better behaved. Eggs would then decay less rapidly with age, and members of the life form could live longer. But that possibility only applies for other-worldly life forms that do not have a merger in their ancestry. Humans are stuck with their eukaryotic ancestry, and are unlikely to evolve a non-destructive meiosis in the near future. If we evolve to live longer it will be by other means. Gamete preservation, as we saw earlier in this chapter, may be one way round the problem. Alternatively, we could make some evolutionary changes in our existing reproductive system. Natural selection could probably favour improved chromosomal surveillance, or some other method of keeping tighter order on the chromosomal dance-floor. Then we could evolve to live longer. A new Methuselah – or rather, his female contemporaries – would need some way of keeping the eggs fresh for longer. It might be done technologically, by freezing. Or it might be evolved, either by staying with our existing destructive meiosis and supervising it more closely, or by evolving a completely different, non-destructive meiosis, perhaps in a gender-free life-form.

I should pause to say I am not at all sure that natural selection really is favouring a slowing down of ageing, and a longer generation time, even in post-industrial human societies. The rate of ageing is evolutionarily tuned to provide the best trade-off between youth and age, in the quality of our bodies. Why our bodies fall to bits in old age is, in the theory of evolution, another way of asking why our bodies are so superbly tuned in youth. Natural selection could lengthen our lives, and improve our bodies at the age of seventy, but it would be at the price of inferior performance at the age of twenty. The direction in which the trade-off is being shifted in modern wealthy societies is a delicate question. I have mentioned the familiar pressure to accumulate more knowledge, for instance in medical education; but there are also pressures in the opposite direction. You do not have to look hard to see a winner-takes-all competition for various attributes, such as beauty or athletic performance, that are closely related to youthfulness. Maybe the winners in these contests are of no Darwinian interest. They may be so odd, or so obsessive, that they do not reproduce at all, or not to any exceptional extent. Alternatively, maybe the cult of youth is not so odd, and natural selection favours superior performance in youth, rather than age, in many areas of human activity. In a modelling, sporting or television career, you could probably gain by tuning up your youthful performance at the expense of a little more decay later: the former would be more of a benefit than the latter would be a cost. In an information-age career you might do better to shift the trade-off in the other direction. There may be conflicting forces in different subcultures, and our future evolution will depend on the net result of the conflict. It is easy to see the advantage of slowing down the ageing process, but in Darwinian terms this does not prove that natural selection favours slower ageing. The slower ageing has to be paid for by decreased quality in youth, and when you realize that the whole argument becomes less certain. It may be the youths, rather than the wrinklies, among our descendants who benefit, over evolutionary time, from the technological improvement of our error rates. Those descendants may still be rationed to three score years and ten, but as youths they will break our athletic records, surpass us in beauty, and outperform us in music and mathematics.

We might evolve information-age mental abilities by learning better instead of for longer. The way we learn our native languages, as young children, is impressive when compared with the learning procedures that we use to learn second languages later in life – and that we also use to learn scientific and computer skills. Later in life we use general learning techniques, such as trial and error, and rote memorization. In contrast, the word 'learning' is almost inadequate to describe the way we acquire our native language – it is more like growing an organ, such as an arm or a leg. The reason, as readers of Steven Pinker's delightful book *The Language Instinct* will appreciate, is that we are programmed to learn much of the structure of our grammar. The programming is probably genetically expensive: there is probably a set of genes concerned only with learning, or internally teaching us, language. We can also use rote memorization to learn a language later in life, but this is by a general learning mechanism that can be used to learn almost anything – a language, names of important people, postage stamp designs. Learning something this way is genetically cheap, in that it does not take extra genes to acquire a new skill, but the bargain price in terms of genes means that the learning technique is crude.

Konrad Lorenz compared the brain mechanisms that underlie learning to an 'innate schoolmarm'. Once we are mature, we learn all sorts of new skills and knowledge, but mainly from one generalist schoolmarm, who teaches all the lessons and knows nothing in particular about any of the subjects. When young, we learn our native language using a specialist schoolmarm, who knows how to teach language, but we soon put her into early retirement. We could learn other skills much more efficiently if we had special internal schoolmarms for them too, but each specialist would need her own piece of DNA code. We could then learn computer skills, or the pricing of financial derivatives, as we learn our native language – with our brains prompting us in the right direction. We could, therefore, evolve to be more complex in terms of our mental abilities either by using some extra genes to code for superior learning skills, or by persisting with our existing abilities and extending our lives – or by both. The future of complexity, in either case, is an extension of what has been going on in human evolution for the past few million years. It is more of the same.

This line of thought may sound familiar. The scale of nature, or the 'great chain of being' (which we met in Chapter 2), is one of the deep ideas of Western thought: the 'great chain' is the set of living creatures, ordered from the simplest to the most complex, in our anthropocentric view of things. The chain would once have had amoebae, or microscopic wheel animalcules, at one end, and led through corals and hydra, worms and insects, fish and frogs to human beings at the other end. Darwin conceptually shattered the chain. The 'great chain of being' was replaced by the 'great tree of life', and in the tree of life there is an equally long history behind a modern bacterium and behind a modern human being. Also, the range of modern species may be ordered in many different ways across the tips of the tree. But that is not my business here. What I am reminded of is that the theoreticians of the past doubted whether the chain from amoebae to humans could be complete. From humans, they reasoned, a further chain, perhaps equal in length to the chain of observable creatures on Earth, would link *Homo sapiens* and God via an ascending hierarchy of angels. Some scholars even worked out a complete pseudo-Linnaean classification of the angels. The historian of the 'great chain of being', Arthur O. Lovejoy, described how the angels in the suprahuman realm differed from one another mainly in their increasing mental powers, culminating in the omniscience of God. The angelic hierarchy was not a matter of sticking a set of wings on one set of angels, and some other useful organ on another set. Gross anatomical variation belongs in the Earthly links of the chain: it is the mental ranges of the chain that are found in the celestial ecosystems. Angels tend to use their brain-power to worship God, rather than to add value in some heavenly information society;

but this need not detain us. Nor am I sure that angelic life forms are coded in DNA, though all analogy with the rest of the chain suggests it is. But whatever the coding chemistry of the angels is, their extra mental powers will not appear in a vacuum; they will require genes, or gene equivalents in some alternative coding molecule, to inform the learning process behind each mental skill. Earthly life has a range of gene numbers from near 10^1 to 10^5. The theologians of the 'great chain of being' liked to argue that humans are equidistant between the simplest life and God. We might deduce that the angelic hierarchy ranges from about 10^5 to 10^{10} genes, with omniscience being coded for by 10^{10} genes.

For some readers, all this will sound too pedestrian: the descriptive natural history of a scientist of limited horizons. I recall how one scientist sneered at another who 'crawls along the frontiers of knowledge with a magnifying glass'. Perhaps I may be permitted some minor speculations in the neglected subject of angelic scholarship. The speculations, indeed, are equally applicable to any life form of superhuman complexity.

It may be sex, Captain – but not as we know it

Superhuman life forms will be coded by more genes than us, or more gene equivalents if they do not use genes and DNA. They may benefit from anti-error techniques even better than ours. In Earthly life, it looks as if the mutation rate and correction efficiency of DNA has been roughly constant since the evolution of bacteria. An error rate of 10^{-10} per letter (after proofreading and repair) is probably some kind of optimum for DNA. We have already looked at how error may be prevented or corrected in the future by gamete freezing and gene therapy. Another idea is that, elsewhere in the Universe, some life form may use a replicating molecule with a lower error rate than DNA. I am too poor a chemist to know whether this is possible, but if it is then it could support more complex life than has evolved on Earth. What I want to think about here, however, is the third mechanism. Could natural selection be made to purge mutational error more efficiently?

In Earthly life, the evolution of sex – and Mendelian sex in particular – was the big breakthrough that improved the efficiency with which natural selection removes mutations. If angels do persist with DNA, and have not perfected gene therapy or something like it, they will certainly need to use sex. Indeed they will need it more than we do, because of their extra genes. Theologians have not all agreed that angels do use sex, but I am not being all that controversial in suggesting they do. Milton, for instance, narrates a long question-and-answer session that took place between Adam and the angel Raphael in the Garden of Eden just before the Fall. Raphael brought things to a diplomatic conclusion when Adam started to ask the kind of questions we are interested in here:

> 'Love not the Heav'nly spirits, and how their love
> Express they, by looks only, or do they mix
> Irradiance, virtual or immediate touch?'
>   To whom the Angel with a smile that glowed
> Celestial rosy red, love's proper hue,
> Answered: 'let it suffice thee that thou know'st
> Us happy, and without love no happiness.'

Raphael then spread his wings and soared back up to heaven. He and Adam never did pick up the conversation, because Satan (disguised as a serpent) intervened and supplied both Adam and Eve with a – how shall we say? – competing system of knowledge. It is uncertain what Adam would have asked next, but if I were him I should want to know whether the angels had evolved a method of sex that concentrated error more powerfully than does the Earthly method.

Earthly sex can improve the efficiency of natural selection. It acts to concentrate a parent's mutations in some of its offspring, but the concentration is not all that impressive. Consider, for instance, a clonal life form in which an average parent makes one mistake per cloned offspring: a parent who produces eight offspring makes eight copying mistakes in all. The life form is unsustainable; it will be destroyed by mutation. What Earthly life does, with Mendelian sex, is redistribute the eight errors among the eight offspring. Mendel's demon sits in the parent and tosses a coin over each gene to decide whether it will be allowed into each offspring. Each gene, including each bad gene, has a half chance of making it into a gamete: four of the gametes on average inherit a harmful mutation and four will be free of error. When the gametes of two such parents are combined to form eight offspring, on average two offspring are free of error, four have one mutation, and two have two mutations. This is an improvement on cloning. It concentrates two mutations in two of the offspring, and changes an impossible system, in which none of the offspring is error-free, into a possible one, in which two of them are error-free. This is summarized in Table 5.

Table 5

Method of reproduction	Number of mutations per parent	Number of offspring with mutations	Number of mutation-free offspring
Cloning	8	8	0
Earthly sex	8	6	2
Angelic sex	8	?	?

I wonder whether in heaven they could concentrate the errors better than us. They might have some method that took all the mutations from seven of the offspring and loaded them into one, producing seven error-free offspring and one scapegoat with eight errors. By way of analogy, God managed to load all human sin into one scapegoat individual, and purge all the moral error of a species in one death. The angelic sex column of the table would then look like this:

Angelic sex	8	1	7

This would be a clear improvement on Mendelian sex. Mendelian sex still took six of the offspring – three-quarters of them – to purge the parent's mutations. The coin-tossing procedure of Mendelism has only a limited power to concentrate error. If all the offspring have the same number of errors, as they roughly do with clonal reproduction, Mendelian sex concentrates the errors. It produces the random distribution. But the random distribution is as far as you can go with coin-tossing. You cannot produce a distribution with seven error-free offspring and one scapegoat. If you toss your coin more furiously, you simply reshuffle an already random distribution; you do not make it more concentrated.

If the error could be concentrated into fewer victims, it would take less death to purge it. Or more error could be purged by the same number of victims, and gene numbers and biological complexity could increase. A life form of almost indefinite complexity could sustain itself if it could redistribute its errors drastically. An angelic brood of eight baby angels could survive with an average of 1000 errors per offspring if angelic sex acted to take 1000 errors out of seven of them, and pile them into the unfortunate offspring number eight. Number eight would expire beneath a load of 8000 errors, but there are still seven error-free offspring to carry the system forwards despite the huge error rate.

A method of sex that concentrates error better than Mendelian sex does have theoretical appeal. Real angels may have come up with it somehow, and mere human reason is an uncertain guide to what they do. But I suspect that such a method is impossible in heaven or Earth. To concentrate errors better than Mendelian sex does, the genetic mechanism would have to recognize which genes were bad copies and which genes were good copies. Many biological mechanisms do indeed recognize the good and the bad copies of genes: proofreading and repair enzymes are examples. But what they do is *correct* errors. They replace the bad copy of a gene by a good copy within the organism. In general, error can be removed either by correction or death. If an error has been detected, it will usually be much cheaper to mend it than to leave the organism to die prematurely. Errors that have been detected within an organism will be corrected if possible. When it is time for sexual reproduction, the only errors left are the ones that have not been detected – errors that have escaped the vigilance of the DNA copyeditors and proofreaders. What are we to do with the undetected errors? They can be eliminated only by natural selection. We should aim to make selection as efficient as possible, producing maximum error elimination for minimum death. We do this by concentrating the errors as much as possible. Ideally we should concentrate all the errors into one scapegoat, but we cannot do this because we do not know which genes to pick on and put in the scapegoat. The best we can do is the random distribution, and that is what life achieves with sex and Mendelism.

Life forms that are more complex than us, therefore, will continue to use sex, unless they have a system of inheritance that is so error-proof that mutations are not a problem. They will also use a kind of sex that behaves in the same coin-tossing way as Mendelian sex does in us. God was able to concentrate all the moral error of the human species into one individual because God can distinguish moral good from evil. If God had not known which was which, it would not have been possible to load all the evil into one sacrificial victim. This leaves only one question. If God could recognize moral error, why purge it in death, albeit one highly efficient death, rather than correcting it with some divine moral repair enzyme? Do not look at me to volunteer to ask prickly old Raphael about that.

Once sex had evolved, the efficiency of natural selection was improved further in Earthly life by the operation of the mating market. Females, in gendered Earthly life, tend to prefer males with error-free genes. By choosing these males, rather than any old male of random genetic quality, a female can improve the quality of her offspring. A frog princess can appreciate the croaks of a male frog, and she will kiss him only if he croaks the croak of accurately copied genes. The genes have become one of the idols of the Earthly mating marketplace, and natural selection has come to act against bad genes with extra force. But it may be that the force could be made stronger. Could the mating, or marriage, markets of more complex life forms discriminate more efficiently against genetic error? One possibility is that the quality-detecting mechanisms could be superior. In fact it is not known how accurately we and other Earthly life forms can measure the gene quality of potential sexual partners. But the mechanisms discovered so far – frog croaks, for instance, and peacocks' tails, and the cha-cha dance steps of the fruit fly – could well be open to improvement. Any mechanism is fallible, because males of low gene quality will (if they can) evolve to fake the voice, apparel or dance steps of their rivals. The advantage of faking will persist in future life forms, but they may evolve some more foolproof method of measuring gene quality than anything life has come up with so far. They might use the results of direct DNA tests. They would have their DNA defects listed in an anonymous, numbered database, and could allow a potential partner to take a look. I am less sure, however, whether the glossy magazines of the time will run features on the romantic experience of 'looking at his DNA profile'.

Pair bonding and the extensive parental care that we go in for probably compromise the antimutational power of human mate choice. Once a high-quality male, or female, has been paired off, his or her availability to the remainder of the population is at least reduced. Parental care is part of the way we create complexity, because parents pass information on to the offspring by non-genetic as well as genetic means. But a life form will not evolve parental care if the genes of the caring parent are not in the offspring. Care by two parents, and pair bonding between those parents, therefore inherently, not just contingently, compromises the genetic powers of mate choice. In a species in which a male and a female simply meet to fertilize the eggs, and the female then lays her eggs and moves on, an individual with exceptionally well-copied genes can be chosen again and again by more than one member of the opposite sex. Future reproductive technologies may make it possible to unbundle pairing, copulating, breeding and parenting, but I doubt whether things will change enough for human mate choice to become more focused on gene quality than it is now. However, some unrelated, more complex life form elsewhere in the Universe might not use pair bonding and parental care like we do, and their uncompromised mating market might favour individuals with high-quality genes more efficiently than ours does.

Mate choice could also be liberated from the constraints of gender. In an ungendered life form, as we considered in Chapter 6, anyone can breed with anyone else; you are not limited to half the rest of the population. There is still an advantage to combining your genes with a partner of high genetic quality, just as in a gendered species. But the members of a species that lacks gender can be even more demanding about whom they breed with, because of the doubled pool of potential partners. Earthling life is gendered, but this will probably prove to be a freakish condition in life as a whole in the Universe.

The absence of gender in the angels may be one reason for their superior biological complexity. I am not sure that angels lack gender (it is another question that Adam could have embarrassed Raphael with), but we have several pointers. European paintings illustrate a large number of angels, in many ecological roles – flying, worshipping, singing and playing musical instruments, delivering flowers and messages, joining in battles on the Christian side – and they all lack visible sex differences. And then we have the findings of the sixteenth- and seventeenth-century witch-hunters. The whole theological research programme was enormous, and I cannot do justice to it, but it yielded incidental results that are relevant to our purposes. The devil (a fallen angel, remember) was found to have sexual intercourse with both he-witches, to whom the devil appeared as a succubus, and she-witches, to whom he appeared as an incubus. The reproductive consequences – such as the birth of Martin Luther, according to the papists – are a complicated subject and may be bestial perversions that tell us little about the practices of unfallen angels. But the results as a whole do suggest that angelic sex is unconstrained by gender.

A final possibility is that sex could be made less of a package deal. We Mendelian creatures have to take all or none of the genes of another individual. We breed by combining all the twenty-three chromosomes of one individual with all the twenty-three chromosomes of another. One suitor may have good disease-resistance genes on chromosome 6 but poor body-building genes on chromosome 12; another may be the other way round. You have to take all or none of a suitor's genes, at least for any one offspring. It would be better if you could take the good chromosome 6 from one of them, and the good chromosome 12 from the other. We cannot do that, but angels may be able to. Angels with twenty-three chromosomes could use twenty-three different courtship displays to reveal the copying accuracy of the genes on each of their chromosomes. After the display for chromosome 1, they produce a sperm that contains chromosome 1, and you can take it or leave it depending how impressed you are by the display. Over time you take in twenty-three different sperm, from up to twenty-three different individuals. You take the best copy of each of the twenty-three, rather than one set that has the best average for the twenty-three as a whole. I imagine there is a problem because an angel with a good chromosome 1 but a bad chromosome 2 may try to sneak his chromosome 2 into the chromosome 1 sperm; but some mechanism should have been able to evolve to deal with that.

The angels in heaven may not have improved on the molecules of inheritance that life on Earth has used for 3500 million years, or on the coin-tossing of Mendel's demon that eukaryotic life has used for 2000 million years. But their mating displays may be more elaborate, designed to show off the quality of distinct gene subsets in turn, and designed to interest all the members of their species, not just those of the opposite sex. What do their displays look like? I only wish I knew. Maybe they are white-knuckle aerobatic displays – dive-bombs, looping-the-loop, free-falls that are broken only just above the ground – and only an angel with the most symmetric wings can bring it off. The evidence from painting, however, shows that angels, like frogs, prefer vocal display, and particularly singing. Singing seems to be a remarkably general accurate indicator of gene quality. One human being in the third century, St Cecilia, could sing like an angel. She was genetically so fussy that she would not allow anyone, even her husband, in her bed, and preserved her virginity for a supernatural lover. Maybe it will be changes in the mating market that provide the next breakthrough in the history of complexity, and enable life to evolve towards the gods.

The rivers of paradise

I have so far mainly been extrapolating from human evolution to think about more complex life forms. This led me, like many others before, to think about life forms that are brainier than we are. But that may not be the way forward. I can imagine that humans now would prefer (and benefit from) some extra genes to code for disease resistance rather than for musical, linguistic or information-technological skills. The wings of angels are suggestive in another way: maybe we could benefit from some whole new organs. Here I want to make a distinction. When we think about genes that code for something extra, we should ask how much that extra facility would cost to use. Wings would probably cost us – they and their muscles have to be grown – and they have to be integrated in a body designed to walk, not fly. There would be many trade-offs in anatomical organization and energy budgets between our existing systems and wings. Indeed the reason we lack wings is probably not that we cannot evolve them, but that they would be ruinous to own. I therefore doubt that extra genes would be used to code for extra organs, or at least for organs on the scale of wings.

Genes for disease resistance are another matter. They would be silent and cost almost nothing, in terms of energy, most of the time. They would be switched on only when they were needed. Many people would now benefit from a gene for HIV resistance, even if the gene were expensive to use. The reason why we in fact do not have more disease-resistance genes is that they would be evolutionarily short-lived. The HIV resistance gene would have been useless fifty years ago, and some different genes will be needed in fifty years' time. Any gene that did defend us successfully against

HIV would tend to cause its own obsolescence as HIV was eliminated, and the gene would then rapidly mutate away. I believe that this explains why we do not have more disease-resistance genes now, and it also has a further implication. If life could drive its mutation rate down, it could load up on genes that are only rarely useful. The genes would be switched on and off, depending on the prevailing plague. If superior mutation-fighting methods allow life in the future to use more genes, the future life forms will probably use the genes to code for occasional needs rather than extra permanent organs like wings.

We could use extra genes to code for wings, if they were switched on only when appropriate. I do not mean that one of our descendants, trapped on a cliff edge and with enemies closing in from the land, would suddenly switch on the wing genes, grow wings and fly off. I have in mind another idea, which I learned from W. D. Hamilton. Sixty thousand genes are needed to code for a human being or a bird, and 20,000 to code for an oak tree or a lobster. One hundred and fifty thousand genes should code for all four. The 150,000 genes would not be needed to build a monster that combined features of all four: this would raise the trade-off and integration problems in a hopelessly insoluble form. What the extra genes provide is the opportunity to choose among very different life forms. At some embryonic stage, our 150,000-gene creature could assess its environment and see where the most promising opportunities lay. If the niche for oak trees was relatively unoccupied, it could commit to this form and grow up as a tree that produces acorns and photosynthesizes. If the sea bottom seemed to be underexploited, it would grow claws, eight legs and a spring-action tail, and watch out for lobster-pots. The embryo would pick the adult life form that promised the best future reproductive return, unconstrained by the form of its parents. The embryo would be a complex creature in itself, because it has to assess all the environments and the opportunities on offer. The life form might start as an assessor-larval form and then undergo a metamorphosis as it switched on the genes of its chosen adult form. All the unused genes would simply be switched off, perhaps until the next generation. The flexible life form could grow up as a lobster in one generation, but produce a brood of birds and oak trees if the sea level went down and the shores dried up. In the lobster generation, 20,000 of the 150,000 genes would be switched on, and the other 130,000 genes switched off. This life form would enjoy great flexibilities that are not open to us, and the flexibility could have its advantages. The flexible 150,000-gene form would not, on average, do worse than its equivalent non-flexible, fixed species; sometimes it would do better.

Why has a 150,000-gene tetra-species flexi-form not evolved, or even a million-gene multispecies life form? There are two related reasons. One is that humans may be near the limit of gene numbers that Earthly life can have. This, however, is only an objection to a form like the human/oak tree/bird/lobster. A lobster/oak tree species could evolve with fewer genes than a human has now. The second reason is that natural selection is needed to preserve genetic information from mutational decay. If an organ is not used, it is lost over evolutionary time, as it is shot through with mutations. Shrimps and fish that have colonized lightless caves have lost their eyes. Their ancestors contained the genes to code for eyes, but these genes were lost when natural selection ceased to act on them. It takes a massive slaughter of mutants every generation to preserve the genetic information of life on Earth: if the genes are not expressed, and presented to selection, they decay.

Life forms carry some silent genes even with our current mutation rates, but the genes have to be used every few generations if they are not to be lost. Male and female bodies differ, for example, and a human female body contains all the genes to build a male genital apparatus or a prominent Adam's apple, but they are switched off. A male body contains the genes for a female genital apparatus and breasts, but they are switched off. This shows there is no inherent problem in having the codes for multiple forms in the same body. But with our mutation rates we cannot take the trend all that far. There is the same problem as with the resistance gene against a rare disease. The gene has to be used every few years or it will be lost. A gene that works exclusively in testicular development is as likely to be miscopied in a female as in a male body, but natural selection only purges the error via a male corpse. Mutations in specifically masculine genes accumulate as fast as they arise in female bodies. This does not create a problem for us, because the genes in a female body now are as likely to be in a male body as a female body in the next generation, and the errors will then be exposed. But a hundred-generation mother–daughter clone might accumulate enough mutations in its silent testicular genes for the hundredth-generation daughter to be unable to produce fertile sons.

Earthly life has been prevented from evolving reserves of occasionally expressed genes by the destructive force of mutation. But what if that force were relaxed or resisted? If future technology or other-wordly evolution drives down the error rate, or turns up the force of natural selection, life could add to its reserves of DNA, and a female human body could contain the implicit codes for something more exotic than a male human body. Then the future of complexity will not lie with the angels. It will lie in life forms who might be no more intelligent than us, but who have genetic subroutines and stand-by life systems that could be called up as appropriate. After fire and brimstone, our descendants will reinvent themselves as fire-adapted flowers and cover the scorched earth with fresh foliage. After the deluge, they will grow up as fish, and swim safely beneath the waves. And when the last trump sounds, they will express the genes for sinlessness, and their ten-thousand-million-gene vessels of cooperating and error-conquering DNA will glide down the rivers of Paradise into eternity.

Glossary

alga (pl. **algae**) A group of **eukaryotes**, including both single-celled aquatic creatures and seaweed. Examples of algae in this book include *Euglena*, a single-celled inhabitant of fresh water; the green algae, which include both the single-celled aquatic *Chlamydomonas* and multicellular seaweeds such as the sea lettuce *Ulva*; and red algae, which include red seaweed. The Cyanobacteria, a group of **prokaryotes**, used to be called blue-green algae, a term I do not use in this book; *Anabaena* is a relatively well-known example of a cyanobacterium.

amino acid The units of which proteins are made. Twenty different amino acids are found in the proteins of life, and the different properties of different proteins are due to their different amino acid sequences and composition. The amino acid sequence of a protein is coded in the DNA and RNA, via the **genetic code**.

amniotes A subgroup of **vertebrate** animals, consisting of reptiles, birds and mammals. They all use a special kind of egg, called the amniotic egg.

anisogamy The condition in which reproduction is sexual and the male and female **gametes** are very different in size, the sperm (or pollen) being much smaller than the egg. Most complex life forms are anisogamous. Compare with **isogamy**.

Archaea A group of single-celled organisms. One of the two divisions of the **prokaryotes**; the other is the **bacteria**.

arthropods A group of animals, including insects, spiders and crustaceans (such as lobsters and shrimps). They are **invertebrates**, lacking a backbone, and they have a hard outside surface, or exoskeleton.

bacteria A group of single-celled organisms. They are one of the two divisions of the **prokaryotes**; the other is the **Archaea**.

base The informational part of a **nucleotide**, and therefore of **DNA** and **RNA**. There are four kinds of base, symbolized by A, C, G and T in DNA and by A, C, G and U in RNA. A triplet of three consecutive bases codes (see **genetic code**) for one **amino acid** in a **protein**.

carbohydrate One of the main chemicals used by living creatures. Sugars are carbohydrates, and carbohydrates therefore include the main energy sources of life.

chloroplast An **organelle** found in many plant cells. Chloroplasts perform photosynthesis, in which energy and **carbohydrates** are produced from sunlight, water and carbon dioxide (CO_2). Chloroplasts contain a small number of **genes**, separate from the majority of the genes in the **nucleus** of the cell.

chromosome The physical structure that contains the DNA. Chromosomes are sometimes visible with a light microscope. An individual organism contains a number of chromosomes that is characteristic of its **species**; each human being, for instance, contains a **diploid** set of 46 chromosomes, made up of one 23-chromosome set inherited from the father and one 23-chromosome set inherited from the mother. The 23 chromosomes differ in size and other respects and are individually numbered – 1, 2, 3 ... 22 and an X or a Y sex chromosome. The DNA of other species is contained in other numbers of chromosomes. A chromosome is a chemical mix of DNA and various **proteins**, which hold the DNA in shape or otherwise interact with or protect it.

clonal reproduction Reproduction in which the offspring is genetically identical to the parent. Only one parent is needed to produce an offspring. It is contrasted with sexual reproduction, in which two parents mix their genes in the offspring. Inheritance in clonal reproduction does not follow **Mendel's laws**.

coding DNA The part of the DNA that codes for **genes**. Compare **non-coding DNA**.

crossing-over The process in which the two copies of each **chromosome** swap their DNA during the reproduction of an organism. It happens during the special cell division called **meiosis**. It is visible in a light microscope. See also **recombination**.

diploid, diploidy An organism or cell, or the condition, in which there are two copies of each **gene** and each **chromosome**: one from the father and the other from the mother. Complex life forms, including us, are usually diploid, except in our **gametes** which reduce the number of copies of each gene from two to one, during **meiosis**. Compare **haploid**.

DNA The molecule of heredity, and the informational basis of life. It stands for deoxyribonucleic acid. The DNA molecule consists of a sequence of **nucleotides**, and is shaped as a double helix.

duplication (1) The kind of **mutation** in which part of the **DNA** is doubled, such that there are two copies of a **gene** or some other region of the DNA where there was one before. (2) The evolutionary event in which a duplication mutation occurs and is then established in all members of the **species**, usually by natural selection.

enzyme A kind of **protein** that catalyses a metabolic reaction (see **metabolism**). Enzymes are biological catalysts. Most of the proteins in the body are enzymes, and most of the living processes in a body only take place because they are catalysed by enzymes.

eukaryote One of the main two groups of cellular life; **prokaryotes** are the other. Eukaryotes include single-celled microscopic forms and also all visible macroscopic forms. Human beings are eukaryotes, as are insects and corals and flowers. Eukaryotes are built of **eukaryotic cells**.

eukaryotic cell Strictly speaking, a cell in which the DNA is enclosed in a distinct **nucleus**. Compare **prokaryotic cell**. However, eukaryotic cells also differ from prokaryotic cells in many other respects. Eukaryotic cells contain DNA-bearing **organelles** – **mitochondria** and **chloroplasts**. Eukaryotic cells generally contain more DNA, and more **genes**, than prokaryotic cells. This book is more concerned with these other differences than with the strict difference, of the presence or absence of a nucleus.

fertilization The biological term for conception: the combination of egg and sperm to form a **zygote**. Not to be confused with agricultural or horticultural fertilizers.

fluke A kind of parasitic worm, which lives inside its host. Liver flukes, for instance, live parasitically inside the liver.

gamete Sperm or egg in animals, pollen or egg in plants: the single reproductive cells that fuse at fertilization to form the **zygote**. They are genetically **haploid**.

gene Informational unit in the DNA. Crudely speaking, one gene codes for one **protein**, but there are so many exceptions to this statement that the term does not have an agreed formal meaning in biology any more.

gene number The number of different **genes** in an individual organism.

genetic code The relation between the 64 possible triplets of **bases** (or **nucleotides**) in the RNA or DNA and the 20 amino acids used in the **proteins** of life.

genome The whole set of DNA of an individual organism. It can be divided into **coding** and **non-coding** regions.

genome project A project to decipher the DNA sequence of a **species**. The human genome project, for instance, is deciphering the human genome. The expression 'genome sequencing project' is also used.

genotype The combination of two genes in a diploid organism; it can be a homozygote or a heterozygote.

globin Family of proteins, including haemoglobin, that carry oxygen in the body.

haemoglobin Complex protein which carries oxygen in the blood. Comprises four globin proteins.

haploid, haploidy An organism or cell, or the condition, that contains one copy of each gene. Bacteria are haploid, but most complex life forms, including ourselves, are diploid (except in our gametes, which are haploid).

harmful mutation A mutation that lowers the survival and/or reproduction of the organism. Synonyms include bad mutation, deleterious mutation, disadvantageous mutation.

heterozygote Diploid genotype in which the organism inherits different versions of a gene from its two parents. Compare homozygote.

homozygote Diploid genotype in which the organism inherits two identical copies of a gene from its two parents. Compare heterozygote.

hormone A molecular signal that circulates (in the blood) in the body, in multicellular organisms, and controls or influences metabolic processes. A hormone often works by interacting with a receptor molecule in a cell's outer membrane.

hybrid Organism produced by crossing parents belonging to two different species.

invertebrates The group of animals without backbones. They include arthropods, snails, worms, starfish, jellyfish, sea anemones, corals, sponges, and many more. Compare vertebrates: animals are divided into invertebrates and vertebrates, and most – probably over 99 per cent – of animals are invertebrates.

isogamy The condition in which reproduction is sexual and the two gametes are roughly the same size. Some relatively obscure microorganisms are isogamous, but most complex life forms are anisogamous.

jumping gene A piece of DNA that has the ability to copy itself from one place to another within the DNA. More formally called a transposable element.

junk DNA DNA that has no apparent coding, or any other, function and is probably useless. It may be neutral or slightly disadvantageous because of the cost of copying it. Non-coding DNA is often suspected of being junk DNA.

lawbreaker This is not a technical term, but I use it in this book to describe a gene that violates the first of **Mendel's laws**, by being inherited in more than 50 per cent of the offspring.

linkage Two genes that are on the same chromosome are said to be linked. They will be passed on to an offspring together, rather than independently (thus violating the second of **Mendel's laws**) unless **recombination** strikes between them.

meiosis The special cell division that reduces the double gene set of a **diploid** organism to a single **haploid** set and produces the **gametes**. Compare **mitosis**.

membrane (1) The cell membrane in **eukaryotic cells** is the external coat of the cell. It is mainly made of a special kind of fat (or, more exactly, lipid) and can be imagined as not unlike a soap bubble; it also contains proteins that act as **receptor molecules** or in other ways. Some eukaryotic cells, particularly plant cells, contain a solid wall outside the membrane. **Prokaryotic cells** (such as bacteria) do not have a cell membrane, but have a rigid exterior. (2) The nuclear membrane in eukaryotic cells has a similar chemical make-up to the cell membrane but encloses the **nucleus**. (3) Eukaryotic cells also contain various internal membranes.

Mendelian genetics The biological system of inheritance discovered by Gregor Mendel (1822–1884), and that is at least the basis for all our modern understanding of genetics. In the strict sense it is the inheritance method used by **eukaryotes**, with **diploidy, chromosome** pairs, and **recombination**. See also **Mendel's laws**.

Mendel's laws Two laws that describe how **genes** are inherited from parent to offspring in **diploid** creatures such as human beings. A human contains two copies (one from the mother, the other from the father) of each of about 60,000 genes. The first law describes how the two copies of any one of the genes are passed on. One of the two copies is inherited in any one **gamete**, and the two copies both have the same 50 per cent chance of being picked to put in a gamete and so inherited in any one offspring. It is also important in this book that the 50 per cent chance is probabilistic; it is not true, for instance, that the maternal and paternal copies of the gene are alternately put in successive gametes. The second law describes the combination of different genes that is passed from parent to offspring. It states that the different genes are inherited independently. The maternal copy of one gene, for instance, has an equal chance of being passed on with the paternal or the maternal copy of any other gene. The second law is produced by **recombination** and is violated by **linkage** between genes.

metabolism   The set of chemical processes going on in the body that maintain the body and keep it alive: digesting food, respiration, supplying energy to our muscles, thermoregulation, and so on.

mitochondrion (pl. mitochondria)   Rod-shaped **organelle** inside most **eukaryotic cells**. Mitochondria are the sites where fuel is burned in oxygen to release energy. They are descended from formerly free-living **bacteria**, and contain a small number of **genes**, separate from the majority of the genes in the cell **nucleus**.

mitosis   Cell division, in a **eukaryotic cell**.

multicellular   Describes an organism made up of many cells. Multicellular creatures begin life as a single cell, the **zygote**.

mutagen   Something that causes **mutation**. Ultraviolet light is an example.

mutation   Any alteration to the **DNA** sequence of an organism. Mainly arises as a copying mistake, but can also arise as a spontaneous chemical change in a **base**. Mutations are the raw material for evolution. In terms of how **natural selection** acts, they may be **harmful**, **neutral** or advantageous, depending on whether the organism with the mutated sequence survives and reproduces worse, as well as, or better than organisms with the unmutated DNA.

natural selection   The process by which kinds of organisms (or, more exactly, **genes** – see **selfish genes**) that reproduce themselves more than other kinds of organisms (or genes) come to increase in frequency over the generations.

neutral   A neutral **mutation** is one that does not alter the survival or reproductive success of the organism. **Natural selection** does not work on it. Neutral mutations undergo neutral evolution: over time the frequency of neutral mutations wander up and down at random.

niche   The range of ecological resources occupied by a **species**. The full description of a species' niche includes all its biological relationships – the food it eats, the competitors it has to maintain itself against, and so on. It also includes the non-biological, environmental resources it occupies: the range of temperatures it can tolerate, the habitat it lives in, and so on.

non-coding DNA   Regions of the **DNA** that do not code for **genes**. It is sometimes thought to be **neutral**, but the sequences of much noncoding DNA are non-random, suggesting that natural selection operates on it to some extent. However, its evolution is probably more neutral than **coding DNA**, and it may be a reasonable assumption that much of the evolution of non-coding DNA is neutral. Non-coding DNA is often suspected of being **junk**.

nucleotide The unit in the molecular chains that make up **DNA** and **RNA**. Both DNA and RNA consist of a 'backbone' of alternating **sugar** and phosphate molecules, with a **base** attached to the sugar. A nucleotide consists of one base–sugar–phosphate unit. Only the base differs between nucleotides, and the letter symbol of the base can also stand for the nucleotide.

nucleus Distinct region, surrounded by a **membrane**, within a **eukaryotic cell** that contains the DNA. Absent in **prokaryotes**.

oocyte A stage in the development of the female **gamete**, or egg cell. It is the '4*n*' stage (Figure 10, p. 186) in which the cell is held at the beginning of the suspended **meiosis**. The two meiotic cell divisions convert the oocyte into the egg. (Technically, the cell at the 4*n* stage is a 'primary oocyte' and after the first meiotic cell division it becomes a 'secondary oocyte'. But the secondary oocyte is a fleeting stage, and oocyte is often used, including in this book, as a synonym for primary oocyte.)

organelle A distinct structure inside a cell. In this book it is usually used as a collective noun for the two kinds of DNA-bearing organelles, **mitochondria** and **chloroplasts**; but there are other organelles inside cells too.

placenta Structure in the pregnant womb of some mammals, including human beings. It nourishes the fetus, using the maternal blood supply. It also secretes **hormones** into the maternal blood and contains **receptor molecules** that interact with hormones from the mother.

ploidy The number of copies of a unit set of **chromosomes** in an individual, or an individual cell. The unit set of chromosomes in human beings has 23 chromosomes. Humans are **diploid** except in their gametes, which are **haploid**. Other species have higher numbers of chromosome sets: they are **polyploid**.

polymerase An **enzyme** that copies **DNA** (a DNA polymerase) or **RNA** (an RNA polymerase).

polyploid, polyploidy The condition of having more than two copies of a unit set of chromosomes. Compare **haploid** and **diploid**.

prokaryote One of the two largest groups of cellular life; **eukaryotes** are the other. Prokaryotes are mainly single-celled creatures. **Archaea** and **bacteria** are the two groups of prokaryotes. Prokaryotes evolved before eukaryotes, and eukaryotes are evolutionarily derived by a symbiosis from prokaryotes. The cells of prokaryotes lack a distinct nucleus.

prokaryotic cell   Strictly speaking, a cell that lacks a **nucleus**; its DNA lies naked in the cell. Compare with **eukaryotic cell**. However, prokaryotic cells differ from eukaryotic cells in many other respects. Prokaryotic cells lack the DNA-bearing **organelles, mitochondria** and **chloroplasts**. Prokaryotic cells generally have less DNA, and fewer **genes**, than eukaryotic cells. This book is more concerned with these other differences than with the strict difference concerning the presence or absence of a nucleus.

proofreading enzyme   An **enzyme** that spots **mutations** in the copying of DNA and has them corrected.

protein   An important group of molecules in living creatures: they perform most of the metabolic (see **metabolism**) processes in the body and its cells. They consist of chains of **amino acids**. **Enzymes** are proteins.

receptor molecule   A molecule (usually a **protein** or a compound of a **sugar** and a protein) that interacts with some other molecule, such as a **hormone**. In consequence of the interaction, it influences the metabolic processes of the cell or body (see **metabolism**).

recombination   Often referred to in this book as **gene** shuffling. The combinations of genes that an organism inherits from its father and its mother are shuffled (or recombined) when that organism reproduces. Recombination happens between genes on different **chromosomes** because the maternal (or the paternal) chromosomes do not move as a group during **meiosis**. A maternal copy of chromosome 1 is equally likely to end up in a **gamete** with the paternal or the maternal copy of chromosome 2. Recombination happens between different genes on the same chromosome because the two copies of the chromosome physically swap stretches of their DNA, in the process called **crossing-over**. Recombination breaks up **linkage** between genes and is the basis of the second of **Mendel's laws**.

repair enzyme   An **enzyme** that repairs damage or mutations in the DNA.

RNA   In most life, including us, RNA is an intermediary molecule. An RNA molecule is read off a **gene** in the DNA. The RNA contains the codes from which a protein is assembled. In some simple life forms, such as **RNA viruses**, RNA is used instead of DNA as the hereditary molecule. RNA stands for ribonucleic acid.

RNA virus   A group of **viruses** that use RNA as their hereditary molecule. They do not contain DNA. Examples include the flu virus and human immunodeficiency virus (HIV), the agent of AIDS.

selfish gene   A gene that has some attribute enabling it to copy itself better than some alternative gene. (Nothing is implied about the motives of the gene; a gene is a bit of a **DNA** molecule and molecules do not have motives.) The phrase 'selfish gene' is usually a way of talking about **natural selection**. Natural selection favours selfish genes. The idea is particularly useful when thinking about how natural selection works on attributes of the DNA.

species   The basic unit of biological classification. In a formal biological classification, using the classificatory system invented by Linnaeus, a species is named in a binomial of generic and specific name, such as *Homo sapiens*. Human beings are an example of a species. A species if often conceptualized as a group of organisms that can interbreed together. This concept is probably not universally satisfactory, and some things that are formally defined as species may owe their specific status more to relative similarity of appearance than interbreeding. Also, closely related species may be able to interbreed in some conditions, producing **hybrids**.

sugar   Sugars are a whole class of molecules that includes glucose and the common sugar (sucrose) that we eat. Our bodies contain many different molecules that belong to the class of sugars. Sugars are examples of **carbohydrates**.

symbiosis   Intimate relationship between members of two biological species.

transposable element   A region of the DNA that has the ability to copy itself (transpose) into another place within the DNA. These elements have characteristic sequence features. They cause **mutations** when they move. Also less formally called **jumping genes**.

vertebrates   The group of animals with backbones: fish, amphibians, reptiles, birds and mammals. Compare **invertebrates**: animals can be divided into vertebrates and invertebrates.

virus   A group of very simple living things, and the main examples of non-cellular life. Viruses are parasites that live inside living cells. In the dispersal stage, between host cells, a virus consists of a hereditary molecule (either **DNA** or **RNA**) wrapped up in a **protein**. When inside a host cell the protein is discarded and the viral DNA or RNA takes control of the cell. See also **RNA virus**.

zygote   The fertilized egg: the initial, one-cell stage in the life of an organism. It is formed by the fusion of **gametes**; genetically it is **diploid**.

Notes and references

The notes are arranged by page of the main text. The references are fully listed in the bibliography, which is a separate section following the notes. Where possible, more introductory and general references for each chapter precede the notes.

Preface

p. ix 'themes of this book': my two themes are not the only ones in the evolution of complexity. *See* Maynard Smith and Szathmáry (1995, 1999); Dawkins (1982); Kauffman (1993).

p. x 'Maxwell's demon': Maxwell (1871), p. 308 in 3rd edn (1872); Garber et al. (1995), pp. 54–67, 180 (for naming by Thomson); von Baeyer (1998) is a popular book on Maxwell's demon.

Chapter 1

The two books by Maynard Smith and Szathmáry (1995, 1999) are about the evolution of complexity; the second (1999) is written for a broader audience, but they are both eminently readable. Gould (1996) discusses trends from the simple to the complex.

p. 1 'van Leeuwenhoek': Dobell (1932), pp. 110–111 for quotation from 1676 publication, p. 304 for pronunciation; Lovejoy (1936) for the literary reception. The Thomson quote is from *The Seasons*, 'Summer', lines 313–316.

p. 3 Pasteur and the microbe hunters: de Kruif (1927); Latour (1984) for Pasteur's reception; Pouchet (1865); Fenner and Gibbs (1988) for viruses.

p. 3 ubiquitous microbes: Fuhrman (1999), Whitman et al. (1998); Gold (1999); Fredrickson and Onstott (1996); Margulis and Sagan (1986); Gould (1996); Pace (1997) for genetic diversity.

p. 5 'Universal Darwinism': Dawkins (1983); Gould (1989).

p. 5 **Romantic literature and the meaning of life:** de Almeida (1991) and her references.

p. 6 'life is anything that can evolve by natural selection': I think I owe this definition to Richard Dawkins (unpublished). There may be a question of whether we want to count all replicating, selectable entities as alive. Advertising jingles copy themselves into our brains; computer files copy themselves between disks; quantal configurations of electrons probably copy themselves in atoms. Maybe these are alive; maybe not. Nothing I am arguing about turns on the issue. Some may prefer to add to the definition some scope-restricting clauses to rule out replicators that have been invented by human beings, or that otherwise violate our sense of life.

p. 8 **hens:** Butler (1878); the quotation is p. 134 (Ch. 8) of the 1910 edition. Pejorative merely: Medawar and Medawar (1977), p. 169.

p. 9 **test-tube experiments:** Spiegelman (1970); Maynard Smith and Szathmáry (1995). Fifteen-fold increase in copying rate: Mills et al. (1967).

p. 12 'origin of life': fossil chemicals in Greenland: Rosing (1999). Molecules: Doolittle et al. (1996); Feng et al. (1997).

p. 13 'origin of cells' (fossil dates): Schopf (1999).

p. 14 'Archaea': Woese (1998).

p. 14 'Russell Doolittle': Feng et al. (1997).

p. 15 **multicellular bacteria:** Shapiro and Dworkin (1997); Shapiro (1988). *Anabaena*: Haselkorn (1998).

p. 16 **unevolving prokaryotes:** Schopf (1994); I reproduced the relevant pictures in Ridley (1997). The 3500 mya Australian fossils were first convincingly published in 1993 (see Schopf (1999)).

p. 16 'largest known bacterial cells': Angert et al (1993).

p. 17 **prokaryotic and eukaryotic differences:** Margulis (1993); Margulis and Sagan (1986).

p. 19 **origin of eukaryotic cells (fossil dates):** Han and Runnegar (1992) for the Michigan corkscrews; Schopf (1999); Knoll (1992); Vidal (1984). Brocks et al. (1999) for the fats. Kerr (1999b) for oxygen.

p. 21 **origin of animals:** 580-million-year-old Chinese fossils: Xiao et al. (1998). Uncertain 1000-million-year-old worm traces: Seilacher et al. (1998); cf. Kerr (1999a). Molecular date: Wray et al. (1996); Bromham et al. (1998).

p. 22 **Table 1:** for vertebrates: Janvier (1999); Kumar and Hedges

(1998); Doolittle et al. (1996); Feng et al. (1997 – and their discussion of other authorities). For humans: Lewin (1998).

p. 23 **probability of origin of life:** Boyden (1953) and references; Monod (1970); Popper (1974), p. 270. Crick (1981); for Stanley Miller see Deamer and Fleischaker (1994).

p. 26 **reverse probabilities:** Carter (1983); Barrow and Tipler (1986, section 8.6); Gott (1993); Leslie (1996), pp. 241–242.

p. 26 **'Rockefeller':** the story is probably apocryphal. I have also heard it told about Joseph F. Kennedy and an elevator boy, and improved by the idea that he shorted the market rather than merely sold out.

Chapter 2

Bonner (1988) is a general book about the evolution of complexity, and particularly discusses the ecological argument that I use at the end of the chapter.

p. 28 **'complexity':** cell types, Bonner (1988); I include his picture in the first two editions of *Evolution* (Ridley, 1996a, p. 636). Alberts et al. (1997) gives 200, not 120, for the number of cell types in a human being. Life stages: Williams (1966), quoted from p. 45. Information theory: Pringle (1951), quoted from p. 176; Thorpe (1963).

p. 30 **Kauffman** (1993), p. 462.

p. 31 **'some snags':** (1) Non-genetic information: Lorenz (1965). (2) Genes can vary in their complexity; for instance genes may be bigger in some life forms than others. Both factors are in principle measurable, but as yet unmeasured. I suspect they are both really correlated with gene numbers, across the range of life forms. (3) Non-coding, but genetic, information in the DNA. Non-coding DNA is often treated as junk, but it may be informational. (By the way, the paradox that the amount of DNA is not simply correlated with the number of genes, and that there is 'too much' DNA in many organisms, is strictly speaking irrelevant to my argument. It is an interesting related topic.)

p. 32 **gene number estimates:** mainly Bird (1995). Viruses: Levine (1992). Syphilis bacterium: *Science* 281, 375 (1998). *E. coli*: Blattner et al. (1997). Prokaryotes and yeast: Doolittle (1998). Flowers: Somerville and Somerville (1999).

p. 34 gene duplications: Ohno (1970). In the discussion that follows, I omit one classic explanation, the escape from heterozygous advantage. It is explained in my book (Ridley 1996) and other texts.

p. 35 St Peter's cock: Quignard (1986), p. 33.

p. 36 'jumping genes': Capy (1998). Jumping with an RNA intermediate is a way of life much like an RNA virus, only the parasite never leaves the cell. Nee and Maynard Smith (1990) discuss the parallels: the mutation limit of RNA viruses that we meet in Chapter 4 also applies to these intracellular jumping genes, for instance, limiting their size and so how much DNA they can duplicate.

p. 38 'Clough and Auden': Clough, *The latest decalogue*, 11–16; Auden, *Under which lyre*, 154–5 and 162–5. For the yeast–worm comparison: Chevritz et al. (1998).

p. 40 'gavotte': Hamilton (1996), p. 358. The pairing of the chromosomes is a neat way of achieving the correct division: Maynard Smith and Szathmáry (1999), pp. 87–88.

p. 42 history of gene numbers: merger contribution, Gray et al. (1999); Zhang and Hewitt (1996). Gray et al. (p. 1479) point out that the nucleus also contains other bacterial genes that have nothing to do with mitochondrial function. Vertebrates: Martin (1999). Biblical scholars: Lane Fox (1991), pp. 21–23 for the creative merger in Genesis.

p. 45 'Great Chain of Being': Lovejoy (1936); Ruse (1996).

p. 45 'persistent presuppositions': Lovejoy (1936), p. vii.

p. 45 'it is absurd': Barrett et al. (1987), p. B74.

p. 45 'Spencerian revolution': Bowler (1989).

p. 45 'in either of two ways': Wagner (1998).

p. 48 'forces ... often unrelated to complexity': alternatively, changes up in one species force competing species to change up too. This is the case of an 'arms race' or 'escalation'. See Vermeij (1987).

p. 48 complexity: fixed lower limit and expanding range: Maynard Smith (1972); Bonner (1988); Gould (1996).

p. 49 'pioneering': Bonner (1988). Darwin in Barrett et al. (1987), E95–96. Darwin added a section with similar ideas, titled 'On the degree to which the organisation tends to advance', near the end of Chapter 4 in editions 3–6 of *Origin* ...; it is not in the first edition. See also Darwin to Lyell, *Correspondence of Charles Darwin*, vol. 8, pp. 92–95.

p. 53 fossil studies of complexity: McShea (1996); Gould (1996); Saunders et al. (1999).

Chapter 3

Eigen (1992) is a popular book that includes the topic of this chapter. Maynard Smith and Szathmáry (1995) also contains a section about it. The population genetic theory of mutation-selection balance is introduced in most evolution and population genetics texts, including mine (Ridley, 1996a).

p. 56 'Ill fares the land': varied from Goldsmith, *The deserted village*, 51–2. The original line 52 is 'where wealth accumulates and men decay' but it is eminently variable: the version 'where wealth accumulates and minds decay' is, for instance, well known in Oxford and Cambridge colleges.

p. 69 'mistakes can be detected as mismatches': Lindahl (1993), and entries in Kendrew (1994) on proofreading and repair enzymes.

p. 71 'two gene sets in diploid creatures are not used [for repair]': unless in the special case of double-strand break repair: see Lindahl and Wood (1999).

p. 72 'more often of type 2': Wright (1977), Ch. 11; Kondrashov and Crow (1991); Valero et al. (1992). They also reveal that there is more to the theory.

p. 76 genetics of religious belief: a twin study referred to in Avise (1998, p. 166) suggests a genetic influence on religiosity.

p. 77 'half chance that any one error-free individual will be susceptible': I here assume for simplicity that selection acts independently on the two traits. If mutant individuals are more likely to be susceptible, selection acts non-independently, and there is a further dimension to the analysis. Try Ridley (1996a), Ch. 7.

Chapter 4

p. 80 'frontier science': surveys complementary to this chapter include Maynard Smith and Szathmáry (1995); Burt (1995); Drake et al. (1998); Keightley and Eyre-Walker (1999); Lynch et al. (1999); Wuehtrich (1998).

p. 83 Table 3 sources: see below for the various mentioned taxa. Also 'The *C. elegans* sequencing consortium' (CESC, 1998); I added the RNA genes to the percentage coding figure.

p. 84 origin of life, 'RNA world': papers in Deamer and Fleischaker (1994).

p. 84 original error rates: Eigen (1992); Maynard Smith and Szathmáry (1995). The error rates in an artificial replicating RNA system are similar (Ekland and Bartel, 1996).

p. 85 Darwin to Lyell: *Correspondence of Charles Darwin*, vol. 8, p. 93.

p. 87 RNA virus error rates: Domingo and Holland (1994); Drake and Holland (1999). Also Drake (1993).

p. 88 RNA viruses near the limit: Nee and Maynard Smith (1990); Drake (1993); Domingo and Holland (1994); Drake and Holland (1999). The Darwinian connoisseur can enjoy the literature on mutation rate in RNA viruses. Some papers use total mutation rate estimates to argue that the viruses are near the error limit (i.e. most mutations harmful), others notice a similarity between the estimated mutation rate and evolutionary rates, and use molecule clocks to date viral origins (i.e. most mutations neutral, at least on a simple view), and others assume the high mutation rates of viruses give them an advantage against us (i.e. mutation is net advantageous).

p. 89 'Holland's experiment': see Domingo and Holland (1994). Oude Essink et al. (1997) describe HIV strains with more accurate copying.

p. 90 'proofreading and repair enzymes': appropriate entries in Kendrew (1994); Lindahl and Wood (1999).

p. 90 'how many mistakes *E. coli* makes': Drake (1991); Maynard Smith and Szathmáry (1995). The fraction that do harm could be guessed at 10 per cent from the Mukai-style experiments on bacteria, reviewed in Lynch et al. (1999), giving a genomic harmful mutation rate of 0.0002. However, the measuring conditions were undemanding, and that tends to produce a low estimate of the genomic mutation rate (Shabalina et al., 1997); 10 per cent is then a minimum estimate for the fraction of mutations that are harmful. Bacteria have tightly organized DNA, with little or no extravagance or junk (Table 3). Ochman et al. (1999) give mutation rates, based on ostensibly neutral evolution rates, for bacteria that are about ten times lower than Drake's figures, based on observed spontaneous mutation rates. This would permit a more complicated discussion; I have stayed with Drake's figures, for ease of comparison with RNA viruses.

p. 91 DNA stability: Lindahl (1993); Lindahl and Wood (1999).

p. 93 'in many stages': there are two further stages for which we might have possible evidence. One is between RNA viruses and bacteria, for the copying accuracy with DNA but without repair enzymes. DNA viruses use DNA rather than RNA, but lack proofreading and repair enzymes of their own. They have low error rates (Drake, 1991, 1993; Maynard Smith and Szathmáry, 1995). However, they are of little interest to us because they are corrected by their host's enzymes. Mitochondrial DNA may have little repair (see 'mitochondrial genome: animal' in Kendrew, 1994). Their rate of synonymous evolution is ten times faster than nuclear DNA in humans. But the relative numbers of germ-line replications are uncertain. The number of mitochondria in oocytes (Chapter 6) would be compatible with ten times, or more than ten times, as much replication in mitochondrial DNA as in nuclear DNA. In other species, rates of mitochondrial and nuclear DNA evolution are similar (Lynch and Jarrell, 1993). Secondly, viroids (see Levine, 1992, p. 2, who says that the smallest has 240 nucleotides, and 'viroids' in Kendrew, 1994) may represent a stage between the origin of life and RNA viruses, but there is a similar problem with them as with DNA viruses. RNA viruses, unlike viroids and DNA viruses, probably tell us something about the error rate in a stage of real evolution. The host cell lacks enzymes to improve the accuracy of RNA replication because RNA replication does not happen in the normal life of the cell. The host cell does not 'help' the RNA virus to lower its mutation rates in the way it helps a DNA virus.

p. 93 'same set of enzymes': the copying accuracy can still differ because the enzymes are used more carefully. Proofreading can be improved by slowing down the polymerase (Michod, 1995). Human DNA is copied more slowly, per letter, than bacterial DNA (Lewin, 1997, p. 536). But the slower copying might be for some other reason than accuracy. Human DNA is more tightly coiled than bacterial DNA, and is surrounded by a bigger burden of administrative molecules. Maynard Smith and Szathmáry (1995, 1999) stress the difference between the number of replication forks in bacteria and eukaryotes; but it is unclear to me whether it is a cause or consequence of the difference in genome size, and therefore complexity, of bacteria and eukaryotes. There are also suggestions that DNA repair efficiency varies among different forms of life (Page and Holmes, 1998). Lindahl and Wood (1999) discuss an exception: bacteria have more enzymes to repair ultraviolet damage than mammals. Our furry, nocturnal ancestors may have had little use for ultraviolet repair enzymes and we have not recovered them.

p. 93 '33 divisions in women ... 200–600 in men': women have a fixed number of oocytes, all laid down before birth; men produce new sperm throughout their lives, and the number of cell divisions behind a sperm increases with male age. See notes for Chapter 9; Chang et al. (1994); Hurst and Ellegren (1998). Hurst and Ellegren (1998) point out that estimates vary and another figure of 26 germ-line divisions exists for women. However, the uncertainty is not important here. Also, Ochman et al. (1999) suggest that bacterial mutation rates are ten times lower than the Drake figures I have used; their method opens up further possibilities for conjuring with generation lengths, germ-line cell division numbers, different methods of estimating mutation rates, and raw numbers – but there is enough flexibility in the order-of-magnitude arguments for it to remain possible that copying accuracy per letter is constant from bacteria to us.

p. 95 'heat shock protein 90': Rutherford and Lindquist (1998).

p. 96 '*p53*': Lozano and Elledge (2000).

p. 97 'reverse translation': Dawkins (1982), p. 175.

p. 99 Mukai's method: Recent reviews are by Drake et al. (1998), Keightley and Eyre-Walker (1999) and Lynch et al. (1999). Though I introduce it for flies, it has been used for bacteria too. In the literature, note that U refers sometimes to haploid and sometimes to diploid rates; the diploid rate matters for the theory.

p. 100 'Svetlana Shabalina': Shabalina et al. (1997).

p. 102 'soft' measuring conditions: I am following Shabalina et al. (1997); see the reviews of Mukai's method (note for p. 99) for the full modern mess.

p. 102 'independent estimate': in Table 3, column 4 can be guessed from column 3 multiplied by column 6. The figure for column 6 for the fruit fly is a guess from the picture in *Science*, 287 (2000), 2194–5, and a crude adjustment for heterochromatin.

p. 103 'pseudogenes': the numbers are in many sources, including my text (Ridley, 1996a). Kimura (1991) was one of the first to do the sum. Playing with the numbers: (i) generation time – I use a figure of 30 years; the time should be the average age of a parent, for all the children of a parent; (ii) which pseudogene you use; (iii) whether pseudogene evolution is pan-neutral (Ridley, 1996; Page and Holmes, 1998); (iv) whether the mutation rates of pseudogenes are representative for the DNA as a whole.

p. 103 'extract the number': Lynch et al. (1999); Eyre-Walker and Keightley (1999); Keightley and Eyre-Walker (1999); Kondrashov (1995).

Chapter 5

p. 110 'expressed exactly': Maynard Smith (1971); Williams (1975); Bell (1981). The argument makes several assumptions, but the two-fold cost is a solid liability at least in many species.

p. 112 parasite theory: papers to be reprinted in vol. 2, following up Hamilton (1996). See also the special issues of *Science* (vol. 281, no. 5385 (25 September 1998) and the *Journal of Evolutionary Biology*, vol. 12, no. 6 (November 1999) for recent authoritative overviews on sex.

p. 112 Kondrashov (1988).

p. 112 Maynard Smith car analogy: he says this in the CD that accompanies my text (Ridley, 1996a) in North America but is distributed separately from the text in the rest of the world. The analogy is haploid, and none the worse for it.

p. 118 'dwell on': see Lenski et al. (1999) and their references.

p. 119 genome projects and duplicated genes: Goffeau et al. (1996); Blattner et al. (1997); C. *elegans* sequencing consortium (CESC, 1998); Chervitz et al. (1998). But Wagner (2000) suggests that mutations in yeast do little damage, not because of back-ups by duplicate genes but by unrelated genes.

p. 120 haemoglobin thalassaemias: Weatherall and Clegg (1981), pp. 452–459. Thalassaemias help in resistance to malaria, but I am only concerned with the effect in the absence of malaria.

p. 121 myoglobin knock-out: Garry et al. (1998).

p. 121 ecological competition: Crow (1999).

p. 124 'standardwing': Diamond (1998) p. 655.

p. 126 'grey tree frogs': Welch et al. (1998).

p. 127 'stalk-eyed flies': Wilkinson et al. (1998). If you happen to have a copy of my book *Evolution* (Ridley 1996), you can see one of Wilkinson's colour photos of these flies in it (colour plate 4); they are also on the cover of the issue of *Nature* carrying the Wilkinson paper. They are worth looking at.

p. 127 'ingeniously solved': that is, by the escalating damage relation on the account I gave earlier. Steven Siller (unpubl.) has explored the relation between mate choice and Kondrashov's theory further. Mate choice can itself cause escalating damage, and also favours sex by unloading genetic deaths from females into males; thus mate choice can give an advantage to sex partly via, or independently of, the escalating damage relation.

p. 129 'Bacterial sex': Michod (1995), pp. 106–110 and references.

p. 131 'bacteria are mainly clonal': yes, but bacterial sex may be more common than some experts had thought. Bacteria are mainly kept in the lab in comfortable conditions, when sex may not be useful. The fraction of mutations that are harmful decreases in cosy conditions, as we saw in Chapter 4. Maynard Smith et al. (1993) reported indirect evidence suggesting that natural bacteria do swap sex at detectable rates, perhaps because conditions there are less comfortable. Note also that the cost of sex will differ in bacteria from species that produce males; the exact cost will depend on the details. It could be 50 per cent for shuffled genes in the recipient, but zero for other genes and zero for the donor. Sex could become advantageous at much lower mutation rates.

p. 131 '*Masterpiece*': Bell (1981); de Almeida (1991) p. 348 n13 on the expression.

p. 131 'viruses': Pressing and Reanney (1984), further discussed by Nee and Maynard Smith (1990), showed that segmented RNA viruses are generally bigger than unsegmented ones. Segmented viruses have a number of RNA molecules inside their capsule, somewhat like having several 'chromosomes'. Segmentation predisposes the viruses to one kind of recombination. In the other kind of recombination, 'copy choice' or 'template switching', a polymerase jumps from one RNA molecule to another and there is a break inside an RNA molecule. Both segmented and unsegmented RNA viruses recombine in this way (Worobey and Holmes, 1999). Note (i) For obligate sex, with random mating, viruses would have to show the escalating damage relation. Chao (1994) has tried to test for it, inconclusively. I can imagine that viruses, and bacteria, may not show the escalating damage relation. Their DNA is more economically organized, and they may contain less of the back-up mechanisms that arguably result in the escalating damage relation in complex life. (ii) However, viral sex, like bacterial, at least sometimes has a donor and a recipient. This is certainly true for 'copy choice' recombination. If it is more likely that the inferior RNA strand will not be chosen, the escalating damage relation is then needed. It is unclear to me whether an inferior RNA could make itself receptive in the manner that a bacterium can. 'Copy choice' recombination is inherently connected with reproduction, unlike bacterial sex. (iii) Viruses may not pay a two-fold cost of sex. As for bacteria (see note for p. 131), the cost will depend on the details; but viruses do not produce males. Thus, microbial sex is a promising test area, but not for naive tests. DNA viruses are also suggestive. They have low mutation rates (note to p. 93), and come in both single- and double-stranded forms. There are also both single- and double-stranded RNA viruses.

p. 132 J. B. S. Haldane: Hamilton (1996) p. 366.

p. 133 'destroyer, as well as the creator': I mainly have full eukaryotic, Mendelian sex in mind. If we go back to bacteria, it is possible that (i) sex originally evolved because of mate choice; (ii) then gene-swapping sex evolved, and first worked Kondrashov-wise with random mating and the escalating damage relation; (iii) then discriminating courtship was added on. It is even possible that sex owes its existence to mate choice all the way up.

Chapter 6

Margulis (1993) is a general reference on the symbiotic origin of the eukaryotic cell. Recent reviews and introductions on intragenomic conflict include Haig (1997a) and Partridge and Hurst (1998). Maynard Smith and Szathmáry's two books (1995, 1999) each contain two relevant chapters.

p. 141 'selfish gene': Dawkins (1982, 1989). Here I mainly use the word 'selfish' in a narrower sense than Dawkins.

p. 142 '(or mainly reproduced)': I allude to the indirect evidence of occasional recombination in human mitochondria (Awadalla et al., 1999). If it is confirmed, occasional quantitative hedging will be needed in this and the next chapter, but no principles should be affected.

p. 143 'sea lettuce': Whatley (1982), p. 532.

p. 143 '*Chlamydomonas*': Whatley (1982); Hagemann (1976); Armbrust et al. (1993).

p. 144 cannibalized mice mitochondria: Kaneda et al. (1995), describing the observations of Hiraoka and Hirano. Sutovsky et al. (1999) for humans. Also see Eberhard (1980), p. 245, on sea urchins, where the cannibals are mitochondria themselves.

p. 146 'create gender': Cosmides and Tooby (1981); Hoekstra (1987); Hurst and Hamilton (1992).

p. 149 'Freudian conflict': Ridley (1993). The rapid evolution of the sex-determining gene on the Y chromosome is probably due to a similar Freudian race against cytoplasmic feminizers (Ridley, 1996a, p. 194 has references).

p. 149 mitochondria in eggs: Lightowlers et al. (1997).

p. 150 'pine trees': Whatley (1982). Hurst and Hamilton (1992) for the microbes, such as *Paramecium*. 'EAT ME PLEASE': Kaneda et al. (1995, and see Sutovsky et al., 1999). There are yet more topics, such as the mitochondrial purges in somatic cell hybrids – see 'mitochondrial genetics' in Kendrew (1994) and Hurst (1990) on cytoplasmic parasites.

p. 152 arguable flukes: also, (i) following Hurst (1990), it could be that a life form with no organelle DNA would have vertically transmitted intracellular parasites, or symbionts, and gender would evolve to keep them well behaved. How common they would have been at the time is unknown (if they are horizontally transmitted, gender is a waste of time). (ii) I assume complex life does not solve the conflict problem by some other non-gender mechanism, such as the nuclear exchange of *Paramecium*. (iii) I am ignoring other possible theories of anisogamy, such as that of Parker, Baker and Smith, discussed by Dawkins (1989).

p. 156 'The snark was a boojum': Lewis Carroll, *The hunting of the snark*, Fit the third, line 32. Should you meet a boojum, 'you will softly and suddenly vanish away, / and never be met with again' (lines 33–4).

p. 156 '*Reclinomonas*': Lang et al. (1997).

p. 157 all mitochondria have one origin: Gray et al. (1999).

p. 157 'chloroplasts tell a similar story': Martin et al. (1998).

p. 158 'as Jesus put it': quoted in *Matthew* 25: 29.

p. 159 'mean as well as lean': these remarks may apply less strongly to the organelles of plants, or of some plants. Some plant organelles are known to be larger than those of animals, and the extra DNA is junk DNA rather than informational genes. The DNA of plant organelles is still much smaller than their nuclear DNA, and the same shrinking force has probably operated in plants as in other creatures, but perhaps less powerfully, or interacting with some other factor (Gray et al., 1999).

p. 159 mitochondria and disease: Boore (1997); Lightowlers et al. (1997); Wallace (1992, 1997, 1999).

p. 160 5000 nucleotide deletion in skeletal muscle: Nagley and Wei (1998): see esp. Fig. 1(a) on p. 514 though it is (reasonably – see their caption) forced through the origin.

p. 160 deletions and heart disease: Wallace (1992).

p. 161 most mitochondrial disease due to mitochondrial, not nuclear, genes: Turnbull and Lightowlers (1998).

p. 161 puzzle of genes left in organelles: Palmer (1997); 'mitochondrial genetics: higher eukaryotes' in Kendrew (1994); for codes, see Wallace (1999) and Gray et al. (1999).

p. 162 'future eukaryote that managed to dispose': this need not be our descendants, or anything near us. It could be a protozoan whose descendants in the distant future take over from the vertebrates, and establish some new form of complex life on Earth without organelle genes. There may be undiscovered protozoans that lack mitochondrial genes even now. The creatures with gene-free hydrogenosomes look too reduced to be on the high-road to the future, but who knows?

Chapter 7

The general references for Chapter 6 also apply here. This chapter uses the paradigm of selfish genes, even more deeply than Chapter 6. Thus, Dawkins' book *The Selfish Gene* (Dawkins, 1989, 1st edn 1976) is implicit background, and in his later book (Dawkins, 1982) he extended his ideas to intragenomic conflict. Matt Ridley's *The Origin of Virtue* is a popular book that starts with intragenomic conflict (including the material of my Chapter 8) and uses the concept to explain, in part, human politics (Ridley, 1996b).

p. 168 'mitochondria ... reproduce clonally': again (see second note to Chapter 6) this may be quantitatively compromised if Awadalla et al. (1999) are right, but the point of principle should still stand.

p. 170 'all that different': primogeniture is a possible exception. The Napoleonic Code abolished primogeniture in conquered countries. The resulting reduced size and efficiency of farms may have influenced the common agricultural policy in, and attitudes to free trade in the different countries of, the European Community. But many other factors were at work and even if a strong Napoleonic influence is granted it does not seriously compromise my claim for Brno: the main influence of (symbolic) Austerlitz would have been to add $150–200 to the annual grocery bill per EC household. The rhetoric of political disputes would also differ. But life would still not be all that different.

p. 170 **Mendel:** Orel (1996), Ch. 5 for the pea experiments, including pp. 103, 122 for the numbers.

p. 171 'segregation distorter': Lyttle (1993) for most of the facts in this section.

p. 173 'simple assassin': *sd* codes for a deviant nuclear transport protein (Merrill et al., 1999). Some nuclear transport molecules work in mitosis, and chromosomes bind to nuclear pores; but the gene may work by less direct means.

p. 174 'target gene': Lyttle (1993). There are really dozens of versions, varying in how vulnerable they are. The more repeats, the more vulnerable they are.

p. 177 'One such mechanism': Crow (1991); Haig and Grafen (1991). Similar ideas had been discussed before, but uninformed by the multigenic nature of real distorters: see Leigh (1987), pp. 241–245.

p. 178 'if Mendel had studied': some of Mendel's seven pea features shared a chromosome, but they were not closely linked (Orel, 1996), pp. 208–209. Also, it is more exact to say that shuffling requires an odd number of recombinational break points, not just a break point, between two genes.

p. 179 'Mendel's law is reimposed': three complications. (i) The advantage of recombination requires the assassin and safe-conduct genes to be associated. They will be, because the two have helped each other to spread. It is analogous to the requirement in Chapter 5 that the population contain more 01/10 than 00/11 pairs. (ii) The recombinant organism contains assassin target (AT) and victim safe-conduct (VS) sperm; AT commit suicide, but fertility is still 50 per cent; the advantage to recombination is that the VS offspring are superior – their reproductive value is higher. (iii) In fruit flies, males do not recombine; but the argument applies with recombination every other generation, in female bodies.

p. 181 **Machiavelli:** Burckhardt (1929), p. 76. I quote the Penguin edition (1970) of Machiavelli, pp. 401–2, 405, 406.

p. 183 'gene is passed on as a unit during evolution': those who follow the debates about whether the gene is a unit of selection (Williams, 1966; Dawkins, 1982, 1989) will notice here an argument that genes inevitably evolve to be (rather than phenomenologically are) units of selection.

p. 184 'two steps': the two steps are in a sense less puzzling than I imply. Meiosis may be evolutionarily derived from mitosis. In mitosis, the chromosome numbers are doubled before cell division, as they have to be if the daughter cells are not to be homozygous. Mitosis goes $2n \rightarrow 4n \rightarrow$ cell division $\rightarrow 2n$. However, there is still a problem. Meiosis has to get from $2n$ to n: why not do it directly, without reusing the first $2n \rightarrow 4n$ stage? Why not have meiosis II without meiosis I? In theory, the kind of 'sister-killer' genes we are to consider could also work after mitosis if it went $2n \rightarrow$ cell division $\rightarrow n \rightarrow 2n$, and that really would mess up sexual life forms. However, that kind of cell division would cause homozygosity, which would usually be a problem, though not in gametes.

p. 184   Haig and Grafen (1991).

p. 189   'life has become protected': the two steps of meiosis randomize gene information effectively, even maximally; but it leaves some exploitable information, as we see in Chapter 8. Also, a sister-killer gene could still theoretically work if it had sufficiently advanced skills. It would have to count the sites of recombination, and time its kill after the appropriate cell division depending on its position in the DNA: kill the sister cell after the first division if you are upstream; kill it after the second if you are downstream. It is all too easy to underestimate what genes can do, but those look to me like human, rather than genic, skills. We can at least conclude that a sister-killer gene has a far more difficult recognition problem after a two-step than a one-step meiosis.

p. 190   'endonuclease': to be more exact, it should randomize among informational sites. The meiotic hot spots are in regions of the DNA that are gene-rich, which makes sense on all theories of recombination in this book. For data, see Holmquist and Filipski (1994 – look at Box 1 on p. 67); their analysis assumes that the marker mutants are randomly distributed down the DNA. Incidentally, this means that the endonuclease cannot pick a random site simply by diffusion: though it could diffuse and then pick the nearest appropriate site.

p. 190   **Mira** (1998): Shrewd readers may ask whether Haig's ideas provide a second perspective on ploidy. In Chapters 3 and 4 (note to p. 72) I treated it as a mechanism that masked copying error. I have not worked out the relation between that and Haig's ideas.

p. 191   **Haig** (1993a). A one-step meiosis like the theoretical one in Maynard Smith and Szathmáry (1999), p. 88 (Figure 7.2a) is not a problem. However, their stage 2 (Fig. 7.2b) may not have existed. Bacteria 'haploidize' without cell division, by something like gene conversion. Yeasts use their normal two-step meiosis to haploidize after endomitotic diploidization (Maynard Smith and Szathmáry 1995, p. 152). However haploidy is produced, some kind of selfish gene will be favoured; sister-killers are the ones that are favoured if it is produced by cell division. *Paramecium* have something like a two-step meiosis inside a cell, without cell division. It should be needed against assassins, but not against sister-killers in the strict sense. However, we can imagine a gene very like a sister-killer, which kills the other nucleus as the two nuclei are formed within a cell; it is using information very like a sister-killer (it only needs one locus, for instance) and doubling the genes up will help against it much like against a sister-killer in conventional meiosis.

p. 193 'through a life cycle': sometimes called a 'bottle-necked' life cycle, the evolution of which was one of the key events in the evolution of complexity for additional reasons beside the one discussed here (Dawkins, 1982). I can imagine a pre-Mendelian stage in which an individual either clones itself via an egg, or could contribute more or less than 50 per cent of the genes to the offspring, depending on it and its partner's quality: that would complicate the narrative about sister-killer genes. I should add that the relation between meiosis and multicellularity is a big story. I have described it from an animal perspective, but I believe that most kinds of haploid-diploid life cycle could be fitted into a suitably generalized version of the argument I give. See also the note for p. 191. Anisogamy evolves along with multicellularity (Bell, 1981).

p. 193 'Kondrashov's theory ... is indifferent': if crossing-over were at the second division, segregation at the first division would need to be rejigged to make one paternal and one maternal chromosome go into each daughter cell.

p. 195 'not talking about trivialities': Hsu (1998), for figure of less than 1 per cent at birth and (p. 184) a third of miscarriages due to trisomies. The fraction of miscarriages found with chromosomal abnormalities varies from 20 per cent to 70 per cent, depending on the timing of the sample: more of the earlier miscarriages have chromosomal abnormalities. There are more miscarriages earlier, and 60 per cent of all miscarriages may have chromosomal causes. About two-thirds of chromosomal abnormalities in miscarriages are trisomies and triploidies (68 per cent of a study of 8841). That gives my one-third figure. Boué and Boué (1973) inferred that half of human zygotes have chromosomal aneuploidies. For the fraction of human conceptions that are spontaneously aborted see Forbes (1997) and more in Chapter 9 below. Other species may have less trouble with chromosomal abnormalities, perhaps because they breed younger: see Binkert and Schmid (1977).

p. 195 trisomal sex difference: Gardner and Sutherland (1996), pp. 247, 252.

p. 197 other possibilities: here are two. (i) Sperm selection: aneuploidies may be equally frequent in males and females, but those in sperm are selected out. There is (imperfect) evidence of selection against chromosomally abnormal sperm, for instance from in vitro fertilization. I doubt this explanation, because of the effect of maternal age on the frequency of offspring with chromosomal abnormalities. The relative number of sperm and eggs is independent of age. (ii) The prolonged, suspended animation of early-phase meiosis in females makes accidents more likely than in males, who perform gametic cell divisions more continuously through life. The sex difference in how many meiotic cells become gametes would then be irrelevant. However, the total time available for 'accidents' is necessarily much the same in males and females. Also (iii) see Forbes (1997) for maternal screening.

p. 199 'Theory of Justice': the quotations are from p. 21 and p. 120 of the paperback edition; see also pp. 136–137.

Chapter 8

The theory here is another instance of intragenomic conflict and the general references for that topic given for Chapters 6 and 7 apply here.

p. 203 Figure 11: after several sources in Haig (1993b).

p. 204 David Haig: Haig (1993b) is the source for almost all the theory, ideas and evidence in this chapter, and one of the great papers on evolution written in the 1990s.

p. 207 Robert Trivers: Trivers (1974). Dawkins (1989) explains Trivers' work.

p. 209 'hamstrings the sperm that it is not in': Herrmann et al. (1999).

p. 223 'To make the argument concrete': there are many other possibilities. The A gene could be on some maternal cells, the a on others. Cells might express both. The placenta may also contain cells expressing the paternal self-recognition gene, and further arguments could be made about that.

p. 223 gene recognition: Haig (1996).

p. 224 'imprinting': Haig (1997a,b).

p. 225 'former probably explains the latter': Iwasa and Pomiankowski (1999) offer another theory for genes on the X chromosome.

p. 226 'Cooperation ... evolves between genetic relatives': Hamilton (1996); Dawkins (1989).

p. 226 mouse brains: Li et al. (1999). See also the news piece in *Science* 29 May 1999, vol 280, p. 1346.

Chapter 9

A number of books have been written about current and future evolution, particularly human evolution; but they do not overlap much with the material in this and the next chapter (Dobzhansky, 1962; Dyson, 1998; Wills, 1999).

p. 233   'Oh my soul ...': Pindar, *Odes*, I.

p. 235   'cell divisions in the reproductive cell line': Chang et al. (1994); Hurst and Ellegren (1998). The exact numbers are uncertain, but the uncertainty does not affect the argument here. Male sperm production, and DNA copying rate, slow down with age, so there may not be 1120 divisions behind a 60-year-old's sperm; but on the other hand copying error rates may increase with age.

p. 236   'one extreme': The potential power of selection may seem surprisingly high after the analogy with broken cars in Chapter 5. However, that analogy has exhausted its use by now. We need to think about (i) the number of bad genes in a creature, whether they are new mutations or pre-existing bad genes that are inherited, and (ii) a whole population of sexual creatures. In terms of (i), the need is to produce enough offspring who show no net deterioration in bad gene numbers, rather than error-free offspring; sex shuffles all the bad genes. In terms of (ii) it is the population that needs to produce enough offspring with no net deterioration, not every parental pair. Variation for bad gene numbers builds up in the population. Sex helps most powerfully when there is 'truncation' selection; see Kondrashov (1988) and Crow (1970). Kimura and Maruyama (1966) obtain an intermediate result under a range of (non-truncation selection) conditions.

p. 236   'crucial condition': There is also mate choice: sex could help to purge our mutations indirectly, by making mate choice possible.

p. 237   'mutating ourselves to extinction': Leslie (1996) discusses various causes, but overlooks mutation. Lynch et al. (1999) and Kondrashov (1995) point out that it is at least puzzling how we exist with our mutation rates; Muller (1950) argued likewise, but from older figures for the mutation rate, and older ideas about how mutations are purged.

p. 238   'we are accumulating mutant genes at a higher rate': Eyre-Walker and Keightley (1999) for molecules. Also, we have been undergoing regressive evolution in our muscles, stature, and even our brains, for the past few hundred thousand years. With regressive evolution, it is always difficult to distinguish a mutational explanation from a positive advantage, due to trade-offs.

p. 239   'selection may be relaxed': relaxing the force of selection is not the same as the idea discussed in Chapter 6, that natural selection can be stopped by removing genetic variation. In terms of the formal conditions for natural selection (Ridley, 1996a, Ch. 4; also the 'meaning of life' bit in Chapter 1 above), relaxation corresponds to removing variation in fitness, not removing variation. Variation accumulates when fitness differences are flattened out.

p. 239 'birth weight': I included extracts from the classic London study from the 1930s, and the Italian evidence of relaxed selection, in Ridley (1997).

p. 240 'Demography': there are some facts in Kondrashov (1988). Reproductive values are needed in a full analysis. See Grafen (1998).

p. 241 'persistent rather than temporary': Muller (1950): I reproduced the relevant bit in Ridley (1997).

p. 241 evidence of relaxed selection: Post (1971). Post also wrote the 1962 review and the numbers I used are repeated in the 1971 paper.

p. 242 Darwin's eyesight: Browne (1995), p. 238. By the way, for myopia the mutation rate is probably effectively high because it is influenced by large numbers of genes, mutations in any one of which make things worse.

p. 243 'nearly all human reproduction': but it should be noted that contribution to the human genetic future depends not only on numbers of babies, but their reproductive values (Grafen, 1998). It could be that people will contribute to the future more in proportion to their wealth than the numbers of their babies.

p. 244 freak out and mañana genes: Wills (1999) and Nesse and Williams (1994) discuss selective forces in wealthy societies.

p. 246 sperm in zona pellucida: Silver (1998), p. 37.

p. 247 'Almost 60 per cent': Wilkins (1993).

p. 247 'back to sperm and eggs': mutations in oocytes (Perez et al., 2000; see also Mira (1998)). Hostile to sperm: Birkhead et al. (1993); Eberhard (1996).

p. 249 marriage and mortality: Darwin (1871), Part I, Ch. V [1894 edn, p. 139]. Subsequently: Hu and Goldman (1990).

p. 250 breeding reduces longevity, in married humans: Westendorp and Kirkwood (1998). The evidence is correlational, not causal, and cannot convincingly show that breeding makes you die; but there is experimental evidence from other species that it does.

p. 250 'symmetry': Etcoff (1999).

p. 250 'for and against': the most important argument against mate choice is that the death rate goes back up for people whose marriages come to an end, whether through divorce or death of the partner.

However, it is not convincing, for reasons Darwin (1871) discussed.

p. 253   Thomas Henry Huxley: Huxley (1893). I included it in Ridley (1997), where the quote is on p. 396.

p. 253   Darwin: (1871), Part I [1894 edn, p. 134].

p. 254   'equalize family sizes': it could be a moral, or civilized, aim to reduce *average* family sizes in overcrowded countries, but that is another question. Natural selection works on variation in family sizes.

Chapter 10

The general references for Chapter 9 apply here too.

p. 257   'corrects most of the errors': the chapter only discusses technological analogues of enzymes that correct the code itself. Almost all medicine provides an analogy for troubleshooting enzymes such as heat shock protein 90 (hsp90), which corrects the expression of genes without correcting the code. The main evolutionary consequence of medicine is relaxed selection, as discussed in Chapter 9. Another possible consequence is implied by the original paper about hsp 90 (Rutherford and Lindquist 1998); it suggested the enzyme promotes evolvability, by allowing disguised genetic variation to accumulate. Relaxed selection might likewise promote evolvability. Evolvability is a theme for a chapter, not a footnote.

p. 258   'full reproductive cloning': my argument does not apply against cloning on the smaller scale, such as the cloning of an individual's cells, or even an organ, to replace a damaged part. This may become common. It has the advantage that the cloned cells should not trigger a transplantation reaction. Somatic mutations in the cells would need to be dealt with.

p. 260   safety of cloning: my discussion implicitly contradicts Silver (1998), pp. 103–4. He says (p. 103), 'there is no scientific basis for the belief that cloned children will be any more prone to genetic problems than naturally conceived children.' I do not know about the belief; but it has a good, if conditional, scientific basis. The general point about not messing with (possible) adaptations that we do not understand is the theme of Nesse and Williams (1994) 'Darwinian medicine'.

p. 260   'Holding hands ...': words by Ira Gershwin, for music by George Gershwin, *Damsel in Distress* (1937).

p. 260   Gene therapy and 'anti-gene therapy': Silver (1998), p. 232 and elsewhere; Friedman (1999).

p. 260   'replacing the bad genes with good': we need to be sure that the bad genes are indeed bad. Some genetic diseases are caused by genes that are also beneficial. Sickle cell haemoglobin can cause death by anaemia, or protect you from malaria, depending on the conditions. Biologists often speculate about whether the genes that cause diseases might be beneficial in rare conditions, such as the plagues of the past. You might like to know, before consenting to gene surgery, that the gene you are going to lose happens to confer resistance to the Black Death.

p. 261   'evolutionary scale': germ-line gene therapy is more controversial than somatic applications. The same evolutionary consequences could probably be purchased more cheaply if we use germ-line rather than somatic gene therapy, but either way we could in theory have a population of people with their genetic errors fixed. Gamete freezing necessarily affects only the germ line, because gametes *are* the germ line. If you freeze your gametes, you reduce your mutation rate, preventing mutations that naturally would have occurred.

p. 262   'Jewish populations': Kleinman (1992).

p. 262   tests for Down's syndrome: when the parent decides to abort, we have a cultural force of selection in favour of fair meiosis, to add to the force of natural selection discussed in Chapter 7.

p. 264   'virgin descendants': my argument here, and below, means that the new gene technologies can have big evolutionary consequences. Gordon (1999, p. 2024) concluded that 'neither gene transfer nor any of the other emerging reproductive technologies will ever have a significant impact on human evolution.' He, or his shade, may yet want that one back. He had reasoned only about genetic enhancement, as if that were the only way the technology could influence our evolutionary future. My arguments work with 'negative' gene therapy, which prevents or repairs disease. Gordon is not the only person to overlook the importance of harmful mutations in evolution. We have an elaborately evolved apparatus to make those mutations as invisible as possible – Mendelism, sex, gender, things like that. Evolutionary changes in this apparatus can be quite interesting.

p. 268   Konrad Lorenz: Lorenz (1965).

p. 271   'Love not ...': *Paradise Lost*, Book VIII, lines 614–621.

p. 273   'cheaper to mend': throw-away organisms such as RNA viruses are probably an exception; *see* Chapter 4.

p. 273   'The best we can do': Mendelism is optimal in these conditions.

It uses a 50:50 coin. If it used a weighted coin, with any other combination of probabilities, it would concentrate error less powerfully. In Chapter 7, we saw that the coin has to have 50:50 probabilities and be inherently random, to deal with lawbreakers; here it has to have the same probabilities, but could in principle be deterministic and exact. Arguments could also be made about the second law, and how it has been shaped by lawbreakers and copying error.

p. 274   'improved further': females gain by choosing mates with good genes, whether or not mutation clearance is the function of sex. This is conditional narrative, not logical theoretical science. Also, as we saw in Chapter 5, a gene recipient can perform effective mate choice if it 'knows' it is worse than average, but cannot distinguish the quality of one donor from another: but that probably applies more to bacteria, with a donor and a recipient, than to eukaryotes, with females and males who swap genes. A 'donor' would not evolve to donate genes to an inferior recipient, if it was committed to a swap.

p. 275   'witch-hunters': Trevor-Roper (1967), pp. 95–96.

Bibliography

Alberts, B. and 5 co-authors (1997) *Molecular Biology of the Cell*, 4th edn. Garland, New York.

Angert, E. R., Clements, K. D., & Pace, N. R. (1993) 'The largest bacterium.' *Nature* 362, 239–241.

Armbrust, E. V., Ferris, P. J. & Goodenough, U. (1993) 'A mating type-linked gene cluster expressed in chlamydomonas zygotes participates in the uniparental inheritance of the chloroplast genome.' *Cell* 74: 801–811.

Avise, J. C. (1998) *The Genetic Gods*. Harvard University Press, Cambridge, Mass.

Awadalla, P., Eyre-Walker, A., & Maynard Smith, J. (1999) 'Linkage disequilibrium and recombination in hominid mitochondrial DNA.' *Science* 285: 2524–2525.

Barrett, P. H., Gautrey, P. J., Herbert, S., Kohn, D., & Smith, S. (1987) *Charles Darwin's Notebooks, 1836–1844*. Cambridge University Press, Cambridge.

Barrow, J. D., & Tipler, F. J. (1986) *The Anthropic Cosmological Principle*. Oxford University Press, Oxford.

Bell, G. (1981) *The Masterpiece of Nature*. Croom Helm, London.

Binkert, F., & Schmid, W. (1977) 'Pre-implantation embryos of Chinese hamster. I. Incidence of karyotype anomalies in 226 control embryos.' *Mutation Research* 46, 63–76.

Bird, A. (1995) 'Gene number, noise reduction, and biological complexity.' *Trends in Genetics* 11, 94–100.

Birkhead, T. R., Møller, A. P., & Sutherland, W. J. (1993) 'Why do females make it so difficult for males to fertilize their eggs?' *Journal of Theoretical Biology* 11, 51–60.

Blattner, F. R., and 16 co-authors (1997) 'The complete genome sequence of *Escherichia coli* K-12.' *Science* 277, 1453–1474.

Bonner, J. T. (1988) *The Evolution of Complexity*. Princeton University Press, Princeton, NJ.

Boore, J. L. (1997) 'Transmission of mitochondrial DNA – playing favorites?' *Bioessays* 19, 751–753.

Boué, A., & Boué, J. (1973) Anomalies chromosomiques dans les avortements spontanés. In: Boué A. & Thibault C. (eds) *Les accidents chromosomiques de la reproduction*, pp. 29–55. Institut National de la Santé et de la Recherche Médicale, Paris.

Bowler, P. (1989) *The Non-Darwinian Revolution*. Johns Hopkins University Press, Baltimore.

Boyden, A. A. (1953) 'Comparative evolution – with special reference to primitive mechanisms.' *Evolution* 7, 21–30.

Brocks, J. J., Logan, G. A., Buick, R., & Summons, R. E. (1999) 'Archean molecular fossils and the early rise of Eukaryotes.' *Science* 285, 1033–1036.

Bromham, L., Rambaut, A., Fortey, R., Cooper, A., & Penny, D. (1998) 'Testing the Cambrian explosion hypothesis by using a molecular dating technique.' *Proceedings of the National Academy of Sciences USA* 95, 12386–12389.

Browne, J. (1995) *Charles Darwin. Voyaging*. Jonathan Cape, London.

Burkhardt, J. (1929) *The Civilization of the Renaissance in Italy*. Harrap, London.

Burt, A. (1995) 'The evolution of fitness.' *Evolution* 49, 1–8.

Butler, S. (1878) *Life and Habit* [1910 edn published by Jonathan Cape, London].

Capy, P. (1998) 'A plastic genome.' *Nature* 396, 522–523.

Carter, B. (1983) 'The anthropic principle and its implications for biological evolution.' *Philosophical Transactions of the Royal Society of London* series A, 310, 346–363. Also in McCrea W. H. & Rees M. J. (eds) (1983) *The Constants of Nature*. Royal Society, London.

[CESC] The *C. elegans* sequencing consortium (1998) 'Genome sequence of the nematode *C. elegans*: a platform for investigating biology.' *Science* 282, 2012–2018.

Chang, B. H.-J., Shimmin, L. C., Shyue, S.-K., Hewett-Emmett, D., & Li, W.-H. (1994) 'Weak male-driven molecular evolution in rodents.' *Proceedings of the National Academy of Sciences USA* 91, 827–831.

Chao, L. (1994) 'Evolution of genetic exchange in RNA viruses.' In:

Morse S. S. (ed.) *The Evolutionary Biology of Viruses*. Raven Press, New York [or see the similar piece in *Trends in Ecology and Evolution*, 1992].

Chervitz, S. A. and 12 co-authors (1998) 'Comparison of the complete protein sets of worm and yeast: orthology and divergence.' *Science* 282, 2022–2028.

Cosmides, L., & Tooby, J. (1981) 'Cytoplasmic inheritance and intragenomic conflict.' *Journal of Theoretical Biology* 89, 83–129.

Crick, F. H. (1981) *Life Itself*. Macdonald, London.

Crow, J. F. (1970) Genetic loads and the cost of natural selection. In: Kojima J.-I. (ed.) *Mathematical Topics in Population Genetics*, pp. 128–177. Springer, New York.

Crow, J. F. (1991) 'Why is Mendelian segregation so exact?' *Bioessays* 13, 305–312.

Crow, J. F. (1999) 'The omnipresent process of sex.' *Journal of Evolutionary Biology* 12, 1023–1025.

Darwin, C. (1871) *The Descent of Man*. John Murray, London.

Dawkins, R. (1982) *The Extended Phenotype*. Freeman, Oxford. [Paperback edition: Oxford University Press, Oxford.]

Dawkins, R. (1983) 'Universal Darwinism.' In: Bendall D. S. (ed.), *Evolution: From Molecules to Man*, pp. 403–425. Cambridge University Press, Cambridge.

Dawkins, R. (1989) *The Selfish Gene*, 2nd edn. Oxford University Press, Oxford.

de Almeida, H. (1991) *Romantic Medicine and John Keats*. Oxford University Press, New York.

Deamer, D. W., & Fleischaker G. R. (eds) (1994) *The Origins of Life*. Jones & Bartlett, Boston.

de Kruif, P. (1927) *The Microbe Hunters*. Jonathan Cape, London.

Diamond, J. (1998) 'The bird-watcher's guide to paradise' [book review]. *Nature* 395, 655–656.

Dobell, C. (1932) *Antony van Leeuwenhoek and His 'Little Animals'*. Bale, London.

Dobzhansky, T. (1962) *Mankind Evolving*. Yale University Press, New Haven, Conn.

Domingo, E., & Holland, J. J. (1994) 'Mutation rate and rapid evolution of RNA viruses.' In: Morse S. S. (ed.) *Evolutionary Biology of Viruses*. Raven Press, New York.

Doolittle, R. F. (1998) 'Microbial genomes opened up.' *Nature* 392, 339–342.

Doolittle, R. F., Feng, D.-F., Tsang, S., Chi, G., & Little, E. (1996) 'Determining divergence times of the major kingdoms of living organisms with a protein clock.' *Science* 271, 470–477.

Drake, J. W. (1991) 'A constant rate of spontaneous mutation in DNA-based microbes.' *Proceedings of the National Academy of Sciences USA* 88, 7160–7164.

Drake, J. W. (1993) 'Rates of spontaneous mutation among RNA viruses.' *Proceedings of the National Academy of Sciences USA* 90, 4171–4175.

Drake, J. W., & Holland, J. J. (1999) 'Mutation rates among RNA viruses.' *Proceedings of the National Academy of Sciences USA* 96, 13910–13913.

Drake, J. W., Charlesworth, B., Charlesworth, D., & Crow, J. F. (1998) 'Rates of spontaneous mutation.' *Genetics* 148, 1667–1686.

Dyson, F. (1998) *Imagined Worlds.* Harvard University Press, Cambridge, Mass.

Eberhard, W. G. (1980) 'Evolutionary consequences of intracellular organelle competition.' *Quarterly Review of Biology* 55, 231–249.

Eberhard, W. G. (1996) *Female Control.* Princeton University Press, Princeton, NJ.

Eigen, M. (1992) *Steps to Life.* Oxford University Press, Oxford.

Ekland, E. H., & Bartel, D. P. (1996) 'RNA-catalysed RNA polymerisation using nucleoside triphosphates.' *Nature* 382, 373–376.

Etcoff, N. (1999) *Survival of the Prettiest.* Doubleday, New York.

Eyre-Walker, A., & Keightley, P. D. (1999) 'High genomic deleterious mutation rates in hominids.' *Nature* 397, 344–347.

Feng, D.-F., Cho, G., & Doolittle, R. F. (1997) 'Determining divergence times with a protein clock: update and reevaluation.' *Proceedings of the National Academy of Sciences USA* 94, 13028–13033.

Fenner, F., & Gibbs, A. (eds) (1988) *Portraits of Viruses: A History of Virology.* Karger, Basel.

Forbes, L. S. (1997) 'The evolutionary biology of spontaneous abortions in humans.' *Trends in Ecology and Evolution* 12, 446–450.

Frederickson, J. K., & Onstott, T. C. (1996) 'Microbes deep inside the Earth.' *Scientific American* 275 (October), 68–73.

Friedman, T. (ed.) (1999) *The Development of Human Gene Therapy.* Cold Spring Harbor Laboratory Press, New York.

Fuhrman, J. A. (1999) 'Marine viruses and their biogeochemical and ecological effects.' *Nature* 399, 541–548.

Garber, E., Brush, S. G., & Everitt, C. W. F. (eds) (1995) *Maxwell on Heat and Statistical Mechanics.* Associated University Presses, Cranbury, NJ.

Gardner, R. J. M., & Sutherland, G. R. (1996) *Chromosome Abnormalities and Genetic Counselling.* Oxford University Press, Oxford.

Garry, D. J., and 7 co-authors (1998) 'Mice without myoglobin.' *Nature* 395, 905–908.

Goffeau, A. and 15 co-authors (1996) 'Life with 6000 genes.' *Science* 274, 546–567.

Gold, T. (1999) *The Deep Hot Biosphere.* Springer, New York.

Gordon, J. W. (1999) 'Genetic enhancement in humans.' *Science* 283, 2023–2024.

Gott, J. R. (1993) 'Implications of the Copernican principle for our future prospects.' *Nature* 363, 315–319.

Gould, S. J. (1989) *Wonderful Life.* Norton, New York.

Gould, S. J. (1996) *Full House.* Harmony Books, New York [published as *Life's Grandeur* by Cape, London].

Grafen, A. (1998) 'Fertility and labour supply in *Femina economica.' Journal of Theoretical Biology* 194, 429–455.

Gray, M. W., Burger, G., & Lang, B. F. (1999) 'Mitochondrial evolution.' *Science* 283, 1476–1481.

Hagemann, R. (1976) 'Plastid distribution and plastid competition in higher plants.' In: Bücher, T. (ed.) *Genetics and Biogenesis of Mitochondria and Chloroplasts.* North-Holland, Amsterdam.

Haig, D. (1993a) 'Alternatives to meiosis: the unusual genetics of red algae, microsporidia, and others.' *Journal of Theoretical Biology* 163, 15–31.

Haig, D. (1993b) 'Genetic conflicts in human pregnancy.' *Quarterly Review of Biology* 68, 495–532.

Haig, D. (1996) 'Gestational drive and the green-bearded placenta.' *Proceedings of the National Academy of Sciences USA* 93, 6547–6551.

Haig, D. (1997a) 'The social gene.' In: Krebs, J. R. & Davies, N. B. (eds)

Behavioural Ecology: An Evolutionary Approach, 4th edn, pp. 284–304. Blackwell Science, Oxford.

Haig, D. (1997b) 'Parental antagonism, relatedness asymmetries, and genomic imprinting.' *Proceedings of the Royal Society of London* series B, 264, 1657–1662.

Haig, D., & Grafen, A. (1991) 'Genetic scrambling as a defence against meiotic drive.' *Journal of Theoretical Biology* 153, 531–558.

Hamilton, W. D. (1996) *Narrow Roads of Geneland*, vol. 1. Spektrum/W. H. Freeman, Oxford [subsequently published by Oxford University Press].

Han, T.-M., & Runnegar, B. (1992) 'Megascopic eukaryotic algae from the 2.1-billion-year-old Negaunee iron-formation, Michigan.' *Science* 257, 232–235.

Haselkorn, R. (1998) 'How Cyanobacteria count to 10.' *Science* 282, 891–892.

Herrmann, B. G., Koschorz, B., Wertz, K., McLaughlin, K. J., & Kispert, A. (1999) 'A protein kinase encoded by the *t complex distorter* gene causes non-Mendelian inheritance.' *Nature* 402, 141–146.

Hoekstra, R. (1987) 'The evolution of sexes.' In: Stearns, S. C. (ed.) *The Evolution of Sex and Its Consequences*, pp. 59–91. Birkauser, Basel.

Holmquist, G. P., & Filipski, J. (1994) 'Organisation of mutations along the genome: a prime determinant of genome evolution.' *Trends in Ecology and Evolution* 9, 65–69.

Hsu, L. Y. F. (1998) 'Prenatal diagnosis of chromosomal abnormalities through amniocentesis.' In: Milunsky, A. (ed.) *Genetic Disorders and the Fetus.* Johns Hopkins University Press, Baltimore.

Hu, Y., & Goldman, N. (1990) 'Mortality differentials by marital status: an international comparison.' *Demography* 27, 233–250.

Hurst, L. D. (1990) 'Parasite diversity and the evolution of diploidy, multicellularity, and anisogamy.' *Journal of Theoretical Biology* 144, 429–443.

Hurst, L. D., & Ellegren, H. (1998) 'Sex biases in the mutation rate.' *Trends in Genetics* 14, 116–452.

Hurst, L. D., & Hamilton, W. D. (1992) 'Cytoplasmic fusion and the nature of sexes.' *Proceedings of the Royal Society of London* series B, 247, 189–194.

Huxley, T. H. (1893) *Evolution and Ethics*. Macmillan, London. [Extract in Ridley (1997).]

Iwasa, Y., & Pomiankowski, A. (1999) 'Sex specific X chromosome expression caused by genomic imprinting.' *Journal of Theoretical Biology* 197, 487–495.

Janvier, P. (1999) 'Catching the first fish.' *Nature* 402, 21–22.

Kaneda, H., and 5 co-authors (1995) 'Elimination of paternal mitochondrial DNA in intraspecific crosses during early mouse embryogenesis.' *Proceedings of the National Academy of Sciences USA* 92, 4542–4546.

Kauffman, S. A. (1993) *The Origins of Order*. Oxford University Press, New York.

Keightley, P. D., & Eyre-Walker, A. (1999) 'Terumi Mukai and the riddle of deleterious mutation rates.' *Genetics* 153, 515–23.

Kendrew, J. (ed.) (1994) *Encyclopedia of Molecular Biology*. Blackwell Science, Oxford.

Kerr, R. (1999a) 'Earliest animals growing younger?' *Science* 284, 412.

Kerr, R. (1999b) 'Early life thrived despite Earthly travails.' *Science* 284, 2111–2113.

Kimura, M. (1991) 'Recent developments of the neutral theory viewed from the Wrightian tradition of theoretical population genetics.' *Proceedings of the National Academy of Sciences USA* 88, 5969–5973 [reprinted in Ridley (1997)].

Kimura, M., & Maruyama, T. (1966) 'The mutational load with epistatic gene interactions in fitness.' *Genetics* 54, 1337–1351.

Kleinman, M. (1992) 'An alternative program for the prevention of Tay–Sachs disease.' In: Bonné-Tamir, B. & Adam, A. (eds) *Genetic Diversity Among Jews*, 346–348. Oxford University Press, New York.

Knoll, A. H. (1992) 'The early evolution of the eukaryotes: a geological perspective.' *Science* 256, 622–627.

Kondrashov, A. (1988) 'Deleterious mutations and the evolution of sexual reproduction.' *Nature* 336, 435–440.

Kondrashov, A. (1995) 'Contaminations of the genomes by very slightly deleterious mutations. Why are we not dead 100 times over?' *Journal of Theoretical Biology* 175, 583–594.

Kondrashov, A. S., & Crow, J. F. (1991) 'Haploidy or diploidy: which is better?' *Nature* 351, 314–315.

Kumar, S., & Hedges, S. B. (1998) 'A molecular timescale for vertebrate evolution.' *Nature* 392, 917–920.

Lane Fox, R. (1991) *The Unauthorized Version*. Viking Penguin, London [Knopf, New York, in 1992].

Lang, B. F., and 8 co-authors (1997) 'An ancestral mitochondrial DNA resembling a eubacterial genome in miniature.' *Nature* 387, 493–497.

Latour, B. (1984) *Les microbes: guerre et paix suivi de irréductions*. Métailié, Paris.

Leigh, E. G. (1987) 'Ronald Fisher and the development of evolutionary theory. II. Influences of new variation on evolutionary process.' *Oxford Surveys in Evolutionary Biology* 4, 212–263.

Lenski, R. E., Ofria, C., Collier, T. C., & Adami, C. (1999) 'Genome complexity, robustness, and genetic interactions in digital organisms.' *Nature* 400, 661–664.

Leslie, J. (1996) *The End of the World*. Routledge, London.

Levine, A. J. (1992) *Viruses*. Scientific American Library, New York.

Lewin, B. (1997) *Genes VI*. Oxford University Press, New York.

Lewin, R. (1998) *Principles of Human Evolution*. Blackwell Science, Boston, Mass.

Li, L.-L., Keverne, E. B., Aparicio, S. A., Ishino, F., Barton, S. C., & Surani, M. A. (1999) 'Regulation of maternal behavior and offspring growth by paternally expressed *Peg3*.' *Science* 284, 330–333.

Lightowlers, R. N., Chinnery, P. F., Turnbull, D. M., & Howell, N. (1997) 'Mammalian mitochondrial genetics: heredity, heteroplasmy, and disease.' *Trends in Genetics* 13, 450–455.

Lindahl, T. (1993) 'Instability and decay of the primary structure of DNA.' *Nature* 362, 709–715.

Lindahl, T., & Wood, R. D. (1999) 'Quality control by DNA repair.' *Science* 286, 1897–1905.

Lorenz, K. (1965) *Evolution and Modification of Behavior*. University of Chicago Press, Chicago.

Lovejoy, A. O. (1936) *The Great Chain of Being*. Harvard University Press, Cambridge, Mass.

Lozano, G., & Elledge, S. J. (2000) 'p53 sends nucleotides to repair DNA.' *Nature* 404, 24–25.

Lynch, M., & Jarrell, P. E. (1993) 'A method for calibrating molecular clocks and its application to animal mitochondrial DNA.' *Genetics* 135, 1197–1208.

Lynch, M., and 6 co-authors (1999) 'Perspective: spontaneous deleterious mutation.' *Evolution* 53, 645–663.

Lyttle, T. W. (1993) 'Cheaters sometimes prosper: distortion of Mendelian segregation by meiotic drive.' *Trends in Genetics* 9, 205–210.

Margulis, L. (1993) *Symbiosis in Cell Evolution*, 2nd edn. W. H. Freeman, New York.

Margulis, L., & Sagan, D. (1986) *Microcosmos*. Summit Books, New York [paperback (1997) University of California Press, Berkeley, Calif].

Martin, A. P. (1999) 'Increasing genome complexity by gene duplication and the origin of the vertebrates.' *American Naturalist* 154, 111–128.

Martin, W., and 5 co-authors (1998) 'Gene transfer to the nucleus and the evolution of chloroplasts.' *Nature* 393, 162–165.

Maxwell, J. C. (1871) *Theory of Heat*. Longman, London.

Maynard Smith, J. (1971) 'The origin and maintenance of sex.' In: Williams, G. C. (ed.) *Group Selection*. Aldine, Chicago [reprinted in Maynard Smith (1972) and Ridley (1997)].

Maynard Smith, J. (1972) *On Evolution*. Edinburgh University Press, Edinburgh.

Maynard Smith, J., & Szathmáry, E. (1995) *The Major Transitions of Evolution*. W. H. Freeman, Oxford.

Maynard Smith, J., & Szathmáry, E. (1999) *The Origins of Life*. Oxford University Press, Oxford.

Maynard Smith, J., Smith, N. H., O'Rourke, M., & Spratt, B. G. (1993) 'How clonal are bacteria?' *Proceedings of the National Academy of Sciences USA* 90, 4384–4388.

McShea, D. W. (1996) 'Metazoan complexity and evolution: is there a trend?' *Evolution* 50, 497–492.

Medawar, P. B., & Medawar, J. S. (1977) *The Life Science*. Wildwood House, London.

Merrill, C., Bayraktaroglu, L., Kusano, A., & Ganetzky, B. (1999) 'Truncated RanGAP encoded by the *segration distorter* locus of *Drosophila*.' *Science* 283, 1742–1745.

Michod, R. E. (1995) *Eros and Evolution*. Addison-Wesley, Reading, Mass.

Mills, D. R., Peterson, R. L., & Spiegelman, S. (1967) 'An extracellular Darwinian experiment with a self-duplicating nucleic acid molecule.' *Proceedings of the National Academy of Sciences USA* 58, 217–224.

Mira, A. (1998) 'Why is meiosis arrested?' *Journal of Theoretical Biology* 194, 275–287.

Monod, J. (1970) *Le hasard and la nécessité*. Seuil, Paris [translated as *Chance and Necessity*].

Muller, H. J. (1950) 'Our load of mutations.' *American Journal of Human Genetics* 2, 111–176 [partly reprinted in Ridley (1997)].

Nagley, P., & Wei, Y.-H. (1998) 'Ageing and mammalian mitochondrial genetics.' *Trends in Genetics* 14, 513–517.

Nee, S., & Maynard Smith, J. (1990) 'The evolutionary biology of molecular parasites.' *Parasitology* 100, S5–S18.

Nesse, R. M., & Williams, G. C. (1994) *Why We Get Sick: The New Science of Darwinian Medicine*. Times Books, New York [published as *Evolution and Healing* by Weidenfeld & Nicolson, London].

Ochman, H., Elwyn, S., & Moran, N. (1999) 'Calibrating bacterial evolution.' *Proceedings of the National Academy of Sciences USA* 96, 12638–12643.

Ohno, S. (1970) *Evolution by Gene Duplication*. Springer, Berlin.

Orel, V. (1996) *Gregor Mendel*. Oxford University Press, Oxford.

Oude Essink, B. B., Back, N. K., & Berkhout, B. (1997) 'Increased polymerase fidelity of the 3TC-resistant variants of HIV-1 reverse transcriptase.' *Nucleic Acids Research* 25, 3212–3217.

Pace, N. R. (1997) 'A molecular view of microbial diversity and the biosphere.' *Science* 276, 734–740.

Page, R. D. M., & Holmes, E. C. (1998) *Molecular Evolution*. Blackwell Science, Oxford.

Palmer, J. D. (1997) 'Organelle genomes: going, going, gone!' *Science* 275, 790–791.

Partridge, L., & Hurst, L. D. (1998) 'Sex and conflict.' *Science* 281, 2003–2008.

Perez, G. I., Trbovich, A. M., Gosden, R. G., & Tilly, J. L. (2000) 'Mitochondria and the death of oocytes.' *Nature* 403, 500–501.

Pinker, S. (1994) *The Language Instinct*. Morrow, New York.

Popper, K. R. (1974) 'Scientific reduction and the incompleteness of all science.' In: Ayala, F. J. & Dobzhansky, T. (eds) *Studies in the Philosophy of Biology*, pp. 29–284. Macmillan, London.

Post, R. H. (1971) 'Possible cases of relaxed selection in civilized populations.' *Humangenetik* 13, 253–284.

Pouchet, F. A. (1865) *L'Univers; les infiniment grands et les infiniment petits*. Hachette, Paris.

Pressing, J., & Reanney, D. C. (1984) 'Divided genomes and intrinsic noise.' *Journal of Molecular Evolution* 20, 135–146.

Pringle, J. W. S. (1951) 'On the parallel between learning and evolution.' *Behaviour* 3, 90–110.

Quignard, P. (1986) *Le salon du Wurtemberg*. Gallimard, Paris.

Rawls, J. (1972) *A Theory of Justice*. Clarendon Press, Oxford [paperback (1973) Oxford University Press, Oxford].

Ridley, M. [Matt] (1993) *The Red Queen*. Viking, London.

Ridley, M. [Mark] (1996a) *Evolution*, 2nd edn. Blackwell Science, Boston, Mass [with CD; book and CD published separately by Blackwell Science, Oxford].

Ridley, M. [Matt] (1996b) *The Origin of Virtue*. Viking, London.

Ridley, M. [Mark] (ed.) (1997) *Evolution*. Oxford Reader. Oxford University Press, Oxford.

Rosing, M. T. (1999) '13C-depleted carbon microparticles in >3700-Ma sea-floor sedimentary rocks from West Greenland.' *Science* 283, 674–676.

Ruse, M. (1996) *From Monad to Man*. Harvard University Press, Cambridge, Mass.

Rutherford, S. L., & Lindquist, S. (1998) 'Hsp90 as a capacitor for morphological evolution.' *Nature* 396, 336–342.

Saunders, W. B., Work, D. M., & Nikolaeva, S. V. (1999) 'Evolution of complexity in Paleozoic ammonoid sutures.' *Science* 286, 760–763.

Schopf, J. W. (1994) 'Disparate rates, differing fates: tempo and mode of evolution changed from the Precambrian to the Phanerozoic.' *Proceedings of the National Academy of Sciences USA* 91, 6735–6742 [reprinted in Ridley (1997)].

Schopf, J. W. (1999) *Cradle of Life*. Princeton University Press, Princeton, NJ.

Seilacher, A., Bose, P. K., & Pflüger, F. (1998) 'Triploblastic animals more than 1 billion years ago: trace fossils from India.' *Science* 282, 80–83.

Shabalina, S. A., Yampolsky, L. Y., and Kondrashov, A. S. (1997) 'Rapid decline of fitness in panmictic populations of *Drosophila melanogaster* maintained under relaxed selection.' *Proceedings of the National Academy of Sciences USA* 94, 13034–13039.

Shapiro, J. A. (1988) 'Bacteria as multicellular organisms.' *Scientific American* 258 (June), 62–69.

Shapiro, J. A., & Dworkin, M. (eds) (1997) *Bacteria as Multicellular Organisms*. Oxford University Press, New York.

Silver, L. M. (1998) *Remaking Eden*. Weidenfeld & Nicolson, London.

Somerville, C., & Somerville, S. (1999) 'Plant functional genomics.' *Science* 285, 380–383.

Spiegelman, S. (1970) 'Extracellular evolution of replicating molecules.' In: Schmitt, F. O. (ed.) *The neurosciences: A Second Study Program*, pp. 927–945. Rockefeller University Press, New York.

Sutovsky, P., and 5 co-authors (1999) 'Ubiquitin tag for sperm mitochondria.' *Nature* 402, 371.

Thorpe, W. H. (1963) 'Ethology and the coding problem in germ cell and brain.' *Zeitschrift für Tierpsychologie* 20, 529–551.

Trevor-Roper, H. R. (1967) *Religion, The Reformation, and Social Change*. Macmillan, London.

Trivers, R. L. (1974) 'Parent-offspring conflict.' *American Zoologist* 14, 249–264.

Turnbull, D. M., & Lightowlers, R. N. (1998) 'An essential guide to mtDNA maintenance.' *Nature Genetics* 18, 199–200.

Valero, M., Richerd, S., Perrot, V., & Destombes, C. (1992) 'Evolution of alternation of haploid and diploid phases of the life cycle.' *Trends in Ecology and Evolution* 9, 25–29.

Vermeij, G. (1987) *Evolution and Escalation*. Princeton University Press, Princeton, NJ.

Vidal, G. (1984) 'The oldest eukaryotic cells.' *Scientific American* 250 (February), 32–41.

von Baeyer, H. C. (1998) *Maxwell's Demon: Why Warmth Disperses and Time Passes*. Random House, New York.

Wagner, A. (1998) 'The fate of duplicated genes: loss or new function?' *Bioessays* 20, 785–788.

Wagner, A. (2000). 'Robustness against mutations in genetic networks of yeast.' *Nature Genetics* 24, 355–361.

Wallace, D. C. (1992) 'Mitochondrial genetics: a paradigm for aging and degenerative disease.' *Science* 256, 628–632.

Wallace, D. C. (1997) 'Mitochondrial DNA in aging and disease.' *Scientific American* 277 (August), 22–29.

Wallace, D. C. (1999) 'Mitochondrial diseases in man and mouse.' *Science* 283, 1482–1488.

Weatherall, D. J. & Clegg, J. B. (1981) *The Thalassaemia Syndromes*, 3rd edn. Blackwell Scientific, Oxford.

Welch, A. M., Semlitsch, R. D., & Gerhardt, H. C. (1998) 'Call duration as an indicator of genetic quality in male gray tree frogs.' *Science* 280, 1928–1930.

Westendorp, R. G. J., & Kirkwood, T. B. L. (1998) 'Human longevity at the cost of reproductive success.' *Nature* 396, 743–746.

Whatley, J. (1982) 'Ultrastructure of plastid inheritance: green algae to angiosperms.' *Biological Reviews* 57, 527–571.

Whitman, W. B., Coleman, D. C., & Wiebe, W. J. (1998) 'Prokaryotes: the unseen majority.' *Proceedings of the National Academy of Sciences USA* 95, 6578–6583.

Wilkins, A. S. (1993) *Genetic Analysis of Animal Development*, 2nd edn. Wiley, New York.

Wilkinson, G. S., Presgraves, D. C., and Crymes, L. (1998) 'Male eye span in stalk-eyed flies indicates genetic quality by meiotic drive suppression.' *Nature* 391, 276–279.

Williams, G. C. (1966) *Adaptation and Natural Selection*. Princeton University Press, Princeton, NJ.

Williams, G. C. (1975) *Sex and Evolution*. Princeton University Press, Princeton, NJ.

Wills, C. (1999) *Children of Prometheus*. Perseus Books, Reading, Mass.

Woese, C. R. (1998) 'Default taxonomy: Ernst Mayr's view of the microbial world.' *Proceedings of the National Academy of Sciences USA* 95, 11043–11046.

Worobey, M., & Holmes, E. C. (1999) Evolutionary aspects of recombination in RNA viruses. *Journal of General Virology* 80, 2535–2543.

Wray, G., Levinton, J. S., & Shapiro, L. H. (1996) Molecular evidence for deep Precambrian divergences among Metazoan phyla. *Science* 274, 568–573 [reprinted in Ridley (1997)].

Wright, S. (1977) *Evolution and the Genetics of Populations*, vol. 3.

Experimental Results and Evolutionary Deductions. University of Chicago Press, Chicago.

Wuethrich, B. (1998) 'Why sex? Putting theory to the test?' [news story] *Science* 281, 1980–1982.

Xiao, S., Zhang, Y., & Knoll, A. H. (1998) 'Three-dimensional preservation of algae and animal embryos in a Neoproterozoic phosphorite.' *Nature* 391, 553–558.

Zhang, D.-X., & Hewitt, G. M. (1996) 'Nuclear integrations: challenges for mitochondrial DNA markers.' *Trends in Ecology and Evolution* 11, 247–251.

INDEX